Managing Rural Environments

Andy Owen

Heinemann Educational Publishers
A division of Heinemann Publishers (Oxford) Ltd.
Halley Court, Jordan Hill, Oxford OX2 8EJ

OXFORD   MELBOURNE   AUCKLAND
JOHANNESBURG   BLANTYRE   GABORONE
IBADAN   PORTSMOUTH(NH) USA   CHICAGO

First published 1999

02 01 00 99
10 9 8 7 6 5 4 3 2 1

ISBN 0 435 35237 7

Designed and typeset by **AMR** Ltd

Printed and bound in Spain by Mateu Cromo Artes Graficas SA

Acknowledgements
The authors and publishers would like to thank the following:
Ros Asquith (2.2); © Michelin, Espangna Sheet 446, 16th Edition, 1998 (map 4.22); C Lowe Morna, 'The Power to Change', Zed Books, 1992 (extract 3.25); Countryside Commission (extract 2.4, graphs 2.5, maps 2.11, 2.14, 2.19); Drake, 1994 (map 4.2); F Harrington, 'The ESF and the promotion of rural development under Objective 5b', Social Europe, 2.91 (extract 4.19); Guardian Education, 17.2.98 (extract 2.22); Howard Newby, 'Green and Pleasant Land? Social Change in Rural England', Hutchinson, 1979 (extract 1.3); J Cole and F Cole, 'The Geography of the European Union', Routledge, 1997 (map 4.7, table 4.8, map 4.15, table 4.17, extract 4.18, map 4.20); National Trust, Long Mynd Access Map, 1998 (5.6); Ordnance Survey, crown copyright, (1.7, 1.9, 5.1, 5.7); Ros Asquith, (cartoon 2.2); Shropshire County Council, (table 1.4, map 1.7, map 1.9, graph 1.11, graph 1.17, map 5.1, tables 5.2, 5.3, map 5.7, table 5.8); Shropshire Sustainable Rural Tourism Project (5.10, 5.11, 5.12, 5.13); Shropshire Hills Joint Advisory Committee, Shropshire Hills Advisory Plan, 1996-2006 (extract 5.9); The Guardian, 2.3.98, (extract 1.2), 28.5.98 (graph 1.15; The Independent, 3.6.98 (extract 2.21); The Ramblers Association (advert 2.7); Thin Black Lines, The Cartoonists Syndicate, New York, (cartoon 4.13); Tiffen et al, 'More People, Less Erosion, Environmental Recovery in Kenya', Wiley, 1994 (maps 3.20, 3.24); Tony Binns (ed), 'People and Environment in Africa', Wiley, 1995 (table 3.1, extract 3.18 – from report to the Dept. of Agriculture on 'Soil erosion and land utilisation in the Ukamba Reserve' by Mahler)

Photocredits
Andy Owen (1.1, 1.14, 1.16, 1.18, 2.12, 2.13, 2.17, 4.3, 4.4, 4.6, 4.10, 4.14, 4.21, 5.4); ARTIC (3.28, 3.29, 3.30); David W Jones (2.8); Environmental Images/Chris Westwood (2.16); Oxfam/G Sayer (3.2); PA News/Ben Curtis (2.3); Shropshire County Council (1.10); Skyscan Photo Library/APS (UK) (1.8, 5.5); Still Pictures/Andrew Testa (2.1), Still Pictures/Mark Edwards (3.9)

The publishers have made every effort to trace the copyright holders, but if they have inadvertently overlooked any, they will be pleased to make the necessary arrangements at the first opportunity.

Tel:01865 888058   email:info.he@heinemann.co.uk

Contents

Introduction

What this book is about

This book looks at rural areas, the problems associated with them and how these problems can be resolved. It examines some of the issues facing rural areas in both wealthy and poorer parts of the world.

Geographers study the relationship between people and their environment, whether they live and work in rural or urban environments. Geographers tend to study urban and rural areas separately, as though they are different and mutually exclusive. However, urban and rural are not simple opposites. They have many features in common: both are places where people live and work, and both are changing rapidly. They are also closely linked by flows of people, money, ideas, technology and goods, so changes in one will affect the other.

One example of this link is the movement of people either into or out of the rural environment. In the more economically developed countries (MEDCs) of Europe and America, many people are leaving the cities and the areas that fringe them to move into rural areas. This movement may be temporary; many visit the countryside for rest and recreation. Many people also move permanently and buy a home in a rural environment that is more tranquil than the busy city. This shift in the population balance from town to country is know as 'counterurbanization'.

This book will examine some of the impacts of counterurbanization on the rural environment in the UK and Europe. In the poorer, less economically developed countries (LEDCs) of Africa and Asia the population movement is in the opposite direction. At GCSE you will have studied the impact of this massive rural to urban migration on developing cities. This book will focus on the rural areas that the migrants have left behind.

In this book you will study:

- how Geographers define rural areas
- how conflicts arise over the use of rural space
- how both government and non-governmental organizations are involved in trying to manage the economy and environment of rural places.

When you have studied the five chapters in this book, you should be able to answer the following questions:

1 What are the characteristics of rural places? In what ways are rural areas close to a major town or city different from more isolated rural areas?
2 What are the issues facing rural areas in MEDCs and LEDCs? How are these affected by population movement in or out of those rural areas?
3 How successful are the strategies used by government and non-government organizations to tackle rural problems?

How to use this book

The book is divided into five chapters. Each provides a focus upon rural areas at a different scale and in different parts of the world; more particularly, each chapter provides a focus upon a different kind of issue. In each chapter, text will enable you to read through from beginning to end. You will need to refer to the Figures as you read. The following additional features are threaded into each chapter in order to help you understand its concepts and its content more easily:

Theory boxes

These help to explain geographical processes which are required to understand the issues in each chapter.

Technique boxes

These help you to be able to present, interpret and analyse data in each chapter. This book focuses especially upon essay techniques.

Activities

These help you to interpret data and text, and work with others in understanding different viewpoints on each issue. Some of these are individual, while others are suggested for groups.

At the end of the book, a Summary will enable you to draw together the concepts and so you can reflect upon and revise your learning.

1 South Shropshire: a small-scale study of a rural environment

Introduction

Our image of any place depends on our direct experience of that place, and how carefully that place has been represented to us in the media. The English countryside is viewed in many different ways by different people. For many it is a beautiful and tranquil place: somewhere to go for a holiday or to retire. This 'chocolate box' image of the rural environment is rejected by many people who live in the countryside. For many rural dwellers the countryside may well be beautiful, but it is isolated, lacks services and there are too few jobs. Some argue that the kinds of deprivation found in English cities, such as unemployment, low pay, crime and drug abuse, are just as common in rural areas. This chapter presents an image of one rural environment in England. You will need to judge for yourself whether or not it is fair and impartial. South Shropshire provides an example of the kinds of issues facing many rural parts of the UK and similar issues are examined at a national scale in Chapter 2.

Images of rural life

A minority of the UK population live in rural areas. Many of those who do so commute to work in urban areas. Less than 2 per cent of the working population are now employed directly in agriculture. If most people have comparatively little experience of rural living, how do we form our image of rural life?

Newspapers and television help the urban population to form their image of rural life. But does the media represent rural life accurately? And how should we view the representation of rural traditions such as farming, morris dancing (Figure 1.1) or hunting and shooting? Does the media create positive images of a vital cultural heritage, or does it reinforce negative stereotypes?

A quick glance at the television confirms that most of the news stories and dramas that we watch are located in urban areas. All of the UK's media, national newspapers and television makers are located in urban areas – mostly in south-east England.

◀ **Figure 1.1** The Shropshire Bedlams morris dancing in Bishop's Castle.

Some of our positive images of rural life come from paintings and poems of the eighteenth and nineteenth centuries. The English poet Wordsworth wrote romantic poems about the tranquillity of rural life. He moved to Grasmere in the Lake District in 1799 and used everyday rural scenes and ordinary rural people as the inspiration for his poetry. He believed that God was present in the harmony of nature, and that humankind should therefore feel kinship with nature. Wordsworth's ideology inspired nineteenth-century conservationists. His images of tranquil rural England still colour our own perception of the countryside. However, his view of the ideal rural lifestyle, or Arcadia, is now questioned by academics such as Howard Newby in figure 1.3.

> **Most bizarre was the music. Spice girls aside, most of it was rousing marches - Land of Hope and Glory, Souza's dambusters theme, Rule Britannia - creating a weird affinity between a pro-blood sports rally and the Last Night of the Proms, and seeming to suggest that you can only be truly British if you live in the countryside and like to kill animals.**

◄ **Figure 1.2** Description of music played during the countryside march on London, 1998. From *The Guardian*, 2 March 1998.

> To most inhabitants of rural England ... the countryside supports a serene, idyllic existence, enjoyed by blameless Arcadians happy in their communion with Nature; or alternatively it is a backward and isolated world where boredom vies with boorishness, inducing melancholia and a suspicion of incest. It is not easy to move beyond these images ... perhaps this is because our connection with the mainsprings of rural life is becoming ever more tenuous. Fewer than three people out of every hundred earn their living directly from agriculture, so that even the much larger proportion who live in the countryside are not of it: most are culturally, if not geographically, urban.

◄ **Figure 1.3** From Howard Newby, *Green and Pleasant Land? Social Change in Rural England*, Hutchinson (1979).

1 Consider Figure 1.1. What impression does this image give you? Do images of rural traditions such as morris dancing and hunting create stereotypes of rural life?

2 Read Figures 1.2 and 1.3.
   a) Consider the different attitudes that are represented in these extracts. Do rural traditions such as morris dancing and hunting reinforce negative images of a backward-looking lifestyle or create a positive image of a vital cultural heritage?
   b) Decide whether the words used are neutral, or whether some words convey a positive or negative bias? Look at the use of words such as rustic, idyllic, Arcadian, isolated, traditional, backward.

   c) Why might an estate agent describe a country cottage as 'isolated'? What image is created? How might a rural teenager feel about this, and why?

3 Discuss the results of the survey in Figure 1.4.
   a) Do you find the results surprising? If so, why?
   b) Do these results challenge accepted stereotypes about the countryside? If so, which ones?
   c) Why might government statistics not show fully the extent of rural deprivation?
   d) How might rural deprivation be measured? Suggest possible methods.

DO YOU AGREE OR DISAGREE WITH EACH OF THE STATEMENTS BELOW?

Please tick the appropriate box for each statement. The questionnaire is completely anonymous.

		Agree	Disagree
1	Living in the countryside is adequate compensation for suffering deprivation	2	71
2	Living in a small village is like living in a goldfish bowl	47	25
3	It costs more to live in the countryside	68	7
4	Villages are close-knit communities where everyone helps each other	19	43
5	Access to services is worse in the countryside	71	4
6	We're all in the same boat in the countryside	3	68
7	You never see a poor farmer	6	66
8	Choosing to live in the countryside is becoming restricted to the more affluent	62	10
9	Rural poverty is a thing of the past	0	73
10	Decisions about the countryside are made by people living in towns	63	9
11	Government statistics show where rural deprivation exists	7	58

Source: Conference participants views, 'Hidden from View' conference, Shropshire 9.10.95

▲ **Figure 1.4** People's perception of rural poverty and deprivation (source: Shropshire County Council).

What are rural areas?

Official definitions of what is meant by rural usually use settlement size or population density as their criteria. These definitions vary from one government to another. Within the European Union (EU) there is considerable variation. In Italy and Spain, settlements with a population greater than 10 000 are urban and all others are rural by default. In France the population threshold of an urban settlement is 2000, while in Denmark and Sweden the threshold is only 200 inhabitants. These variations reflect the geographical and cultural differences between societies. Italy and Spain have long histories of urban living and both have internationally important urban centres. Italy, for example, has 50 cities with populations greater than 100 000. Denmark, by contrast, has a sparser population and only four cities in excess of 100 000 people. So smaller settlements would seem comparatively more important to a Dane than to an Italian. However, all of these definitions indicate our urban-centric viewpoint. We define a rural area by what it is not! This is not surprising. European society is generally organized around urban places. Cities are the centres of our governments, our universities, businesses and decision making. No wonder rural people sometimes feel that their viewpoint is ignored.

The rural–urban continuum

Most governments use population size to define what is meant by an urban area. All other areas are rural by default. Since different governments use different criteria, it is impossible to compare urban or rural areas in different countries unless a standard definition is used. The United Nations provides this standard definition by stating that settlements with a population greater than 20 000 are urban.

Even with a standard definition it is sometimes difficult to define where an urban area begins or ends. The distinction between what is truly urban and what is rural is confused by our experience of places that seem to have characteristics common to both. Mechanization of farming has meant that huge numbers of agricultural jobs have been lost since 1945. At the same time, urban dwellers have moved into villages that are within easy commuting distance of cities. Many villages have therefore lost their rural–agricultural function, and have become distant housing estates of the cities, giving them an urban function. There is, therefore, a sliding scale or rural–urban continuum which reflects the increasingly urban function of some rural areas.

It is helpful to distinguish the core of a city from the ring of countryside, towns and villages that immediately surround it and have taken on some of the urban functions. Many people who live in this ring commute daily into work, or use the city for shopping, schools, health and leisure facilities. Beyond the ring is a more sparsely populated rural area. People here may still use the city's facilities, but on a more infrequent and irregular basis.

Urban areas expand outwards and as they do so, they consume what was once open countryside and assimilate it into the edge of the city. The rapid expansion of the city has been particularly common in the USA, Japan and the UK, and is often referred to as suburban sprawl. Sprawl has been a feature of urban growth since the 1950s. It has been made possible by wider car ownership making travel into the city centre for shopping and work easier. In addition, as the city centre has become congested, developers have turned to undeveloped rural sites on the fringe, or edge, of the city for new developments of retailing, office space and factory units. The urban fringe is characterized by change and development. It is a zone in transition, in which rural areas are becoming increasingly urban.

▼ **Figure 1.5** The main features of the daily urban system.

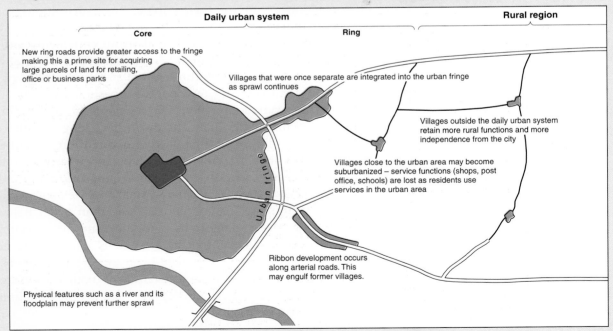

1 Use the model of the daily urban system to make notes on the characteristics of core, ring and rural communities in the UK.

2 In what ways do urban areas (or urban dwellers) consume rural space?

Counterurbanization

In recent decades there has been a gradual change in the perception of city and rural life. Many people still see urban areas as offering greater opportunity than rural areas which are deprived of jobs and entertainment. However, other people now perceive the city as a hostile environment for family life. Noise, air pollution, traffic congestion and crime rates are all *perceived* to be higher than in rural areas. These perceptions can be seen as push–pull factors that influence personal choice about where to live. Coupled to this is greater car ownership and a belief by many that longer journeys to work are an acceptable trade-off against the benefits of a better quality rural life. The result has been a shift in population as people leave urban areas and move into the countryside, a process known as counterurbanization.

Defining counterurbanization

Counterurbanization may be defined by reference to relative change in the population of the daily urban system and the surrounding rural area. Counterurbanization is characterized by the decentralization of people, industry and other urban functions away from cities. The relative populations of urban and rural areas change as a result of demographic factors, such as fewer births, and the migration of people. These population changes result in:

- an overall decline in the population of the daily urban system, where the largest losses may be in the urban core; the urban fringe and ring may also be in decline.
- an increase in the number of people living in the surrounding rural area, as more people migrate from the urban to the rural area.

Counterurbanization is the result of a number of demographic, social and technological changes. These are summarized in Figure 1.6.

▼ **Figure 1.6** Counterurbanization: cause and effect.

Type of change	Cause	Effects on the rural area	Effects on the urban area
Demographic	The size of the average European and US household is in decline. This is a result of: • people having fewer children due to personal choice, greater affluence and the availability of contraceptives • a higher divorce rate • people living longer.	Greater affluence and mobility means there is increased demand for country housing. New building in rural areas can suburbanize a village and is often resisted by the local community.	Divorce and longer life expectancy mean that families become smaller and more dispersed. Some family-sized houses in the urban core are now occupied by single people. Urban population densities fall. Deprived areas of the urban core have higher proportions of single parents and elderly residents who cannot afford to move out.
Social	Greater affluence, the improvement of the transport infrastructure and increased car ownership have made the workforce more mobile.	Accessible rural areas provide homes for daily commuters. More remote rural areas attract commuters who may spend four nights in the city and three in their rural retreat. Commuting increases rural traffic and the demand for housing. House prices rise in village locations as salaried commuters compete in the limited housing market. Locals can no longer afford to buy cottages.	Traffic congestion and associated air pollution (nitrous oxides, low level ozone, dust from break linings, carbon monoxide, etc.) increase on routes in and out of the city.
Technological	New technologies, such as the use of home computers, faxes and the internet, allow workers to operate from a rural home – a development known as out-working or telecottaging.	Rural out-workers have several disadvantages compared to colleagues who work in an office. They are often isolated and lack social contact. They may work on short contracts or at a piece rate, i.e. they are paid for each piece of work they complete. They may be paid less than city workers. Out-workers do, however, have more flexibility in arranging their working hours than those who work in offices.	Office workers in the city lose jobs as work is subcontracted out to more flexible out-workers who have lower overheads and are therefore cheaper.

1 What seem to be the effects of counterurbanization on both city and rural areas? Suggest the main advantages and disadvantages under the headings:

- Environmental
- Social
- Economic.

Introducing South Shropshire

Bishop's Castle is a small, historic market town. It is situated in the south-west of Shropshire, just a few kilometres from the Welsh border, and on the edge of the Shropshire Hills Area of Outstanding Natural Beauty (AONB).

The district of South Shropshire is a rural community. As Figure 1.7 shows, there are no large settlements that can be considered truly urban (for standard definitions of both urban and rural see page 7). Ludlow is the largest settlement in the district with a population of around 11 000. Bishop's Castle has a population of just 1600. Many comprehensive schools have a larger population!

The outlying settlements are tiny hamlets and the population density is extremely low. The scattered or sparse nature of population distribution is an important feature of the area. In one way it helps to strengthen a sense of community, because the isolation from city amenities means that people rely on one another for many social activities and events. But rural isolation can also be a handicap, especially for those who cannot drive. Because there are so few people there is insufficient demand for many services. Facilities that are taken for granted in cities, such as play groups, youth clubs, cinemas and theatres, are very widely scattered here.

▼ **Figure 1.7** The location and settlement distribution of the district of South Shropshire.

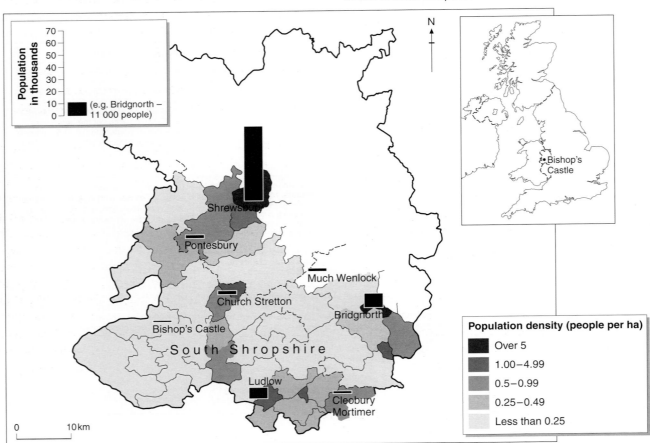

▶ **Figure 1.8** Wenlock Edge, a limestone escarpment that is part of the Shropshire Hills AONB.

The physical environment is characterized by rolling hills, sharp ridges and broad dales. This landscape is the result of three factors:

- the varied geology of the area
- the events at the end of the last ice age when periglacial conditions and melting ice shaped the hills and vales
- centuries of extensive hill farming.

Wenlock Edge in Figure 1.8 is a classic example of how geology can influence the shape of the landscape.

The district is cut in half by the A49 trunk road which links Shrewsbury, to the north with Ludlow, Hereford and then Worcester to the south. Settlements to the east of the A49 are within easy commuting distance of the West Midlands conurbation. Settlements along the A49 are most accessible to tourists visiting the AONB (Figure 5.7, page 73). But settlements to the west of the A49, like Bishop's Castle, are the most inaccessible. All parts of the district are affected to some degree by links with larger settlements in the West Midlands and beyond. Commuters, second home owners, the retired seeking a country home and tourists all make these links which affect different parts of South Shropshire to different degrees.

Population change

The attractive image of rural areas such as Bishop's Castle has made them increasingly popular places to live for those who want to move away from the city, which is known as counterurbanization. Counterurbanization has a number of impacts on the rural area as we have seen, not all of which are beneficial. Two changes are occurring to the age structure and social characteristics of most rural areas generally, and in South Shropshire, specifically:

- Young people aged 18-25 are leaving the rural area in order to attend university or find work.
- Older people aged 40 or more are moving into the rural area.

Older newcomers may have children of their own, in which case the structure of the community is not greatly altered. However, some may have children who have already grown up and left the family home. Others may be retired people looking for a country retreat. In these circumstances the demographic structure of the rural area will gradually change. If this occurs the new community may have quite different needs from the original one. If it has a disproportionate number of either children or people over 65, it will have a high dependency ratio; that is people who are dependent upon social services, such as youth workers or the health service, than a more balanced community.

The area of south-west Shropshire is characterized by both outward and inward migration. In some communities the newcomers have brought children with them. Other communities in the area have attracted many more retired people, as indicated in Figure 1.9.

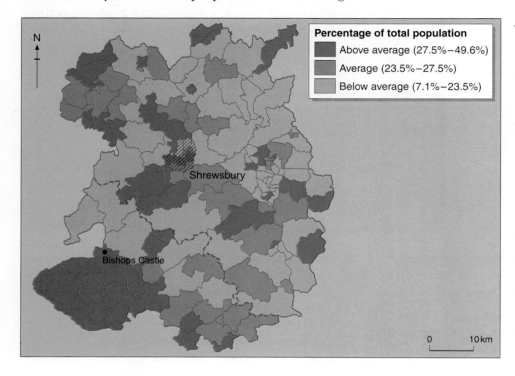

Figure 1.9 Pensioner only households by ward, Shropshire (source: Shropshire County Council).

Rural deprivation

The growth of personal mobility through greater car ownership has meant that accessible rural communities, like those in the eastern part of South Shropshire, have become more densely populated and generally more affluent as commuters move in. Most basic services are retained in such rural areas, because the growing population requires schools, shops and pubs. However, closure is still possible. A village shop selling groceries, for example, finds it difficult to compete against the large out-of-town supermarkets. And commuters may visit the high street bank in their lunch hour, or use a telephone banking system, rather than the small village branch.

Meanwhile, lack of work opportunities in the most isolated or marginal agricultural areas has resulted in economic decline and depopulation. These rural communities are losing essential services such as schools, bus services, shops and post offices. In some of the most remote parts of the UK this deprivation has lead to further decline of the rural population as young people in particular leave the area in order to have better access to education, work and basic services. Such communities may be described as suffering from multiple deprivation: they are deprived of, or lack, a number of features (work and services) that contribute to people's quality of life.

The lack of regular public transport services can further isolate such communities, especially for people who do not drive. For example, over 2700 households (15 per cent of all households) in rural South Shropshire have no car. This may be no more or less than the number of households without a car in many urban areas. However, urban dwellers have more choice of public transport. The difference between choice and constraint is important when considering deprivation.

◀ **Figure 1.10** The library bus serves isolated rural communities in south-west Shropshire. It visits nearly 50 different villages and hamlets every two weeks.

▼ **Figure 1.11** The changing number of key services in rural Shropshire, 1980–90 (source: Shropshire County Council Rural Facilities Surveys).

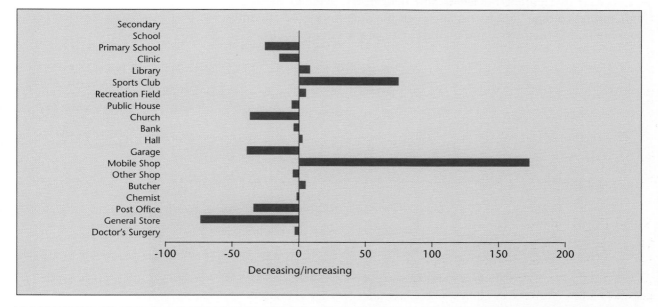

	Shop	Post office	Pub	Village school	Hall/community centre	Bus service
Shropshire	50	50	38	57	33	12
West Midlands	45	46	31	53	30	12
England	39	40	26	51	29	13

▲ **Figure 1.12** Percentage of parishes without a key service.

Bishop's Castle is a community with social interaction and there are very few commuters. The crime rate is low. Shopping trips need planning. The library service is poor and there are few sport opportunities. Insufficient numbers often mean adult education courses are not financially viable.

You have limited mobility if you do not use a car. Because the area looks nice the government does not recognize the extent of rural deprivation. More could be spent on schools, library facilities, youth clubs and buses, subsidized transport.

We bought a house with land so we could have animals. It is in a peaceful location with no next-door neighbours. The children have plenty of space to play in safety and it is a less violent society for them to grow up in. There is less public transport so it is essential to be able to drive. There are few leisure facilities locally so it is necessary to travel long distances, especially where the children's interests and hobbies are concerned.

▶ **Figure 1.13** Responses to a survey about the advantages and disadvantages of living in South Shropshire, a relatively isolated rural community.

1 Use Figure 1.11 to identify the key services that have decreased by the greatest number.

a) Suggest how different groups of people might be affected by the decrease (for example, young families, commuters, those who are retired, those without a car).

b) Suggest reasons for the growth of some key services.

2 Use Figures 1.10, 1.11, 1.12 and 1.13 to assess the view that 'Shropshire is suffering from rural deprivation'. Do you agree that it is?

Changing rural patterns of work: diversification

The distinctive landscape of the Shropshire Hills is partially a product of many centuries of agricultural management. Farmers are responsible directly or indirectly for many landscape features such as those seen in Figure 1.14. Fields and hedgerows, avenues of trees, agricultural cottages and barns, and even the open moors on the hill tops are due to their work and management of the environment.

◀ **Figure 1.14**
The character of the countryside is greatly affected by centuries of settlement and farming.

Rural employment has changed considerably since World War II. Mechanization of agriculture and the sale of small farms has led to many job losses. However, EU subsidies to farmers has meant that farming has continued even in upland areas like the Shropshire Hills where the land is of marginal agricultural value. A number of changes in recent years are threatening the future of farming in these areas. Figure 1.15 indicates the extent of the recent decline of farming. The BSE crisis has caused financial problems for many upland farmers. On top of this, there is the uncertainty over whether the EU will continue to support farming by providing subsidies through the Common Agricultural Policy (CAP). If subsidies are withdrawn, who will manage the rural environment of the Shropshire Hills and what will become of the landscape?

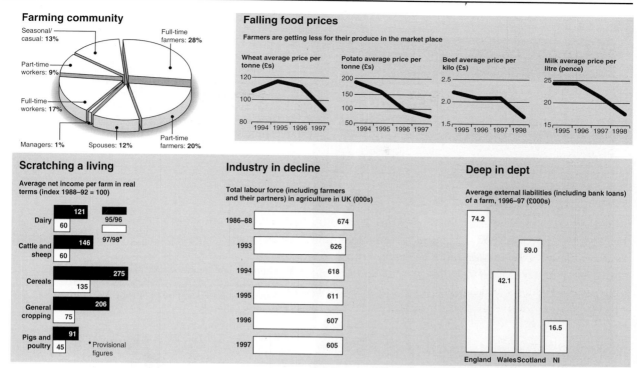

▲ **Figure 1.15** The state of UK agriculture (source: *The Guardian*, 28 May 1998).

The nature of rural employment is having to change. Diversification has meant jobs have been created in services such as tourism, and also re-created in traditional crafts such as furniture making (Figure 1.16). However, the new jobs are often casual or part time. Fruit picking is one example. New rural jobs may also be informal in nature, i.e. without proper regulation of hours, and without proper health and safety checks. In addition to this, self-employment is much more common in rural than in urban areas. An average of 20 per cent of the rural workforce is self-employed, compared to 12 per cent in England as a whole. Many of the self-employed run small businesses which can be at a serious disadvantage compared to larger firms which benefit from economies of scale. Compared to larger firms, small businesses:

- can make only small capital investments
- cannot respond as easily to fluctuating levels of demand by taking on more or less labour
- suffer disproportionately from regulation and red tape.

There is a wider range of incomes for those people who are self-employed. Evidence in Shropshire suggests that the self-employed have a slightly higher weekly mean income than those in employment: £313 opposed to £271 (1993–94).

However, the weekly median wage (half-way value) is lower for the self-employed: £187 rather than £213. This suggests that the lower paid self-employed are worse off than those in employment.

Wages in Shropshire are lower than average UK wages, as can be seen in Figure 1.17.

Rural manufacturing

Bishop's Castle has only two major employers: Ransfords timber yard and the Community College. The two largest factories in the town closed in the early 1990s. One was the Walters trouser factory (Figure 1.18). It had about 40 employees, mainly women. The women who worked here sewed together fabric that had been cut to shape in another branch in Ludlow. It closed because the firm could save money by transporting the fabric to Lithuania in the Baltic, where the garments were sewn and re-imported to the UK by road. The other factory, called Farm Gas, made bio-digesters. It employed about 40 men. After the closures a partnership of local business and public service groups made a bid for aid from central government. Their report described Bishop's Castle as 'a town in terminal decline'. The bid was successful in obtaining some funding. Since then, the area has become eligible for funding from the EU under Objective 5b status (see page 64).

◄ **Figure 1.16** There has been a recent resurgence of interest in traditional rural crafts. Ian is bodging, or turning green wood for chairs. He also makes willow sculptures. Ian worked for 25 years for a major UK firm. He then took a redundancy pay-off and moved to South Shropshire. He is now self-employed and some of his work is in the informal sector.

▼ **Figure 1.17** Average weekly earnings in Shropshire compared to the UK as a whole, 1995 (source: 'Poverty and deprivation in Shropshire', Shropshire County Council).

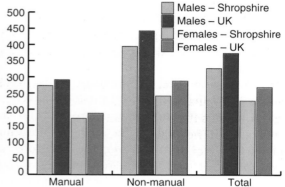

► **Figure 1.18** The old Walters trouser factory has been converted using EU money. It now contains several small industrial units and a community IT centre.

Ideas for further study

Investigate perceptions of rural and urban places. Do different groups of people have different perceptions of what it is like to live in rural and urban places? To what extent does their viewpoint depend on:

- age
- where they have lived or visited
- the media ie TV programmes that they watch, or newspapers that they read.

Summary

- People have different perceptions about the UK rural environment. This is important because:
  - a positive image of the countryside has been one of the reasons for the shift of population from urban to rural places. This in-migration of people has been responsible for some social and economic problems in rural environments.
  - many rural people perceive that the tranquil rural image means that their real needs (for jobs or services) are neglected by governments and other decision makers.
- Recent economic change has threatened economic life in rural areas. Small rural towns such as Bishop's Castle are working environments that may require some assistance to boost their economies. (The issue of recreation in the rural environment is pursued further in Chapter 2. EU assistance for rural economies is examined in Chapter 4.)

References and further reading

Definitions of rurality are discussed in:
Christopher Bull, Peter Daniel and Michael Hopkinson (1984) *The geography of rural resources*. Oliver and Boyd. (especially chapter 1)

Rural issues and the changing nature of rural settlements and services are discussed in:
Howard Newby (1988) *The Countryside in Question*. Hutchinson.
John Chaffey (1994) *A new view of Britain*. Hodder and Stoughton.

Government websites providing statistical information for the UK on issues such as population patterns and change can be found at:
The Office for National Statistics at http://www.ons.gov.uk
The Department of Environment, Transport and the Regions at http://www.detr.gov.uk

Introduction

The UK is a highly urbanized society in which 90 per cent of people live in settlements of 10000 people or more. However, increasing numbers of people are giving up life in the city to move permanently into rural areas. Many others use the countryside for holidays or as a weekend retreat from the pace of city life. The process of counterurbanization is having massive impacts on the UK's rural space and can cause tension and conflict in the countryside between local people and newcomers.

Those who move permanently into the rural environment need housing. House prices may escalate and become too expensive for those locals who are on low incomes. Those who visit the countryside for a holiday or short break may also create conflict. Seasonal traffic congestion on narrow country roads, parking problems, pollution and footpath erosion may all result from giving people more access to the countryside. These problems have to be balanced against economic advantages for rural communities that result from increasing tourism.

This chapter examines just how much access members of the public have to the countryside. It discusses the national and local agencies that are involved in rural recreation and the strategies used to try to ensure that more people have access to the rural environment. For instance, should rambling and tourism be encouraged or controlled in the UK's rural areas?

▼ **Figure 2.1** Many people claim the right to continue country traditions such as hunting, while others are violently opposed.

Conflict in the countryside

There is a tradition of conflict in the UK countryside which social scientists believe has its roots in the historic class system. It exists between country landowners and the urban population or 'townies'. Rural landowners (including farmers) regard themselves as custodians of the countryside. They have a long history of management of the landscape, which includes water and timber resources, and the production of food. Many farming families have worked the same piece of land for several generations. They have a strong attachment to the land and wish to preserve the landscape and rural traditions such as hunting (Figure 2.1), shooting and fishing. While farmers and rural land owners are producers, we may regard urban populations as consumers of the countryside. The urban population wish to enjoy the benefits of country living, whether it be by moving into a country home or by taking part in leisure activities in the countryside. They therefore consume rural space, to the possible resentment of rural people, as is suggested in Figure 2.2.

What is the rural–urban conflict about?

▲ **Figure 2.2** Second home owners can create conflicts in rural areas (source: *The Guardian,* 18 July 1998).

◀ **Figure 2.3** Pro-hunting demonstrators see the anti-hunt lobby as an attack on rural tradition. They believe that parliament does not represent the views of a rural minority. In March 1998 over 250 000 people marched in London to demonstrate their viewpoint on the countryside.

The issues facing rural areas in the UK were brought to the attention of the media by the Countryside Alliance march in London in March 1998 (Figure 2.3). The Alliance is a new and well-organized group with thousands of members and the ability to generate massive publicity and lobby politicians. It is a coalition of different groups who are interested in the countryside, but at its core is the British Field Sport Society who campaign to preserve the right to hunt. The Alliance was formed mainly to protest against an anti-hunting bill put forward by one MP, Michael Foster. Many protesters were concerned solely with demonstrating their right to continue country sports such as hunting, shooting and fishing. However, the march and the resulting media coverage raised other important issues facing rural communities:

- the loss of the greenbelt as more and more houses are built in the countryside
- the decline of the rural economy, the slump in farmers' incomes and the proposed loss of subsidies from Europe for farmers
- the decline of rural services such as village schools, transport and hospitals
- whether or not access should be improved for ramblers and others to enjoy the countryside.

Tourism in the rural economy

Some country landowners resent the invasion of their privacy and worry about trespassers on their land. They resist the demand for the creation of more footpaths and better access to the countryside. So what are the potential benefits of encouraging greater access to the countryside? Tourism can certainly boost the rural economy: it has direct and indirect economic benefits. Jobs are created directly in hotels, restaurants and guest houses. Other businesses may also benefit indirectly from the growth of rural tourism. Farmers can diversify by converting unused barns into holiday lets to provide extra income. Local builders and other businesses benefit indirectly as demand increases for barn conversions.

Recreation and tourism have important economic impacts. Those who stay overnight make the largest contribution to the rural economy, with money spent on food and accommodation. But day visitors also make an impact. In 1996 the Countryside Commission conducted a survey of day visits. Some of their main findings are recorded in Figures 2.4 and 2.5.

▶ **Figure 2.4** From Countryside Commission website.

Between 1994 and 1996 the number of leisure day visits grew by about 10%, from 5.2 billion to 5.7 billion. A similar rate of growth applied to all of the three main types of destination – towns/cities, the countryside and the seaside/coast. In the same period, the average amount spent per trip rose by slightly more than inflation. The combined effect means that expenditure on day visits grew from about £44 billion to about £52 billion, a real growth after inflation of around 13%.

Seven visits in ten were to a town or city (71%), a quarter were to the countryside (26%) and 3% were to the seaside or coast. Two thirds of all visits took place in groups, rather than alone, and a quarter of these visits involved children as well as adults. The average party size was just over 3.

Day visits were an all-year-round activity. The average visit took about 3½ hours; 2½ hours at the destination and 1 hour travelling. The most popular activities were visiting friends and relatives, having a meal or a drink out, walking and shopping. Car was the main mode of transport for nearly three trips in every five (57%), followed by walking (30%). The average amount spent on each trip was £9.10; £10.60 on town trips, £5.10 on countryside trips and £10.20 on seaside trips.

Visits to the three main destinations – town, countryside and seaside – could also be recorded as involving visits to woods/forests or to canals/rivers. Most visits to woods and forests (81%) took place on a countryside trip; the remainder took place on town or seaside trips. Walking was the main activity on three in every five wood/forest visits (61%); for half the visits – more than for any other kind of trip – people also travelled there on foot. The average distance travelled was slightly lower than for countryside trips as a whole, and more of them (two in five) were taken unaccompanied by other people. The average amount spent – £3.15 – was the lowest of all types of trip. In most other respects they were similar to countryside trips.

▼ **Figure 2.5** Number of day visitors to urban, rural and forest areas and average spending, 1996 (source: Countryside Commission).

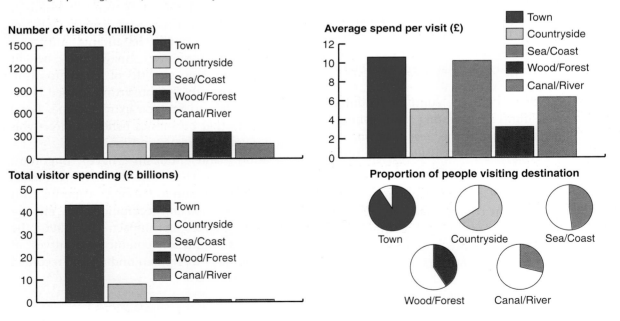

1 Summarize the main differences between visits to urban and visits to rural areas.

2 To what extent do rural communities benefit from urban visitors? How might they also lose out?

3 How might rural economies gain greater benefits from day visitors? Should they encourage more visitors or more spending? Justify your view.

Fifty years of managing the countryside – the Countryside Commission

There is a long history of dispute over the question of whether or not people should be given the right to walk freely through the UK's rural environments. In the early twentieth century rural landowners resisted those who wanted access to their land. A series of mass trespasses by ramblers on privately owned moorland brought huge publicity to those who wanted improved access to the countryside. This eventually resulted in the 1949 National Parks and Access to the Countryside Act. The act established the National Parks Commission (later renamed the Countryside Commission) who then designated the National Parks and Areas of Outstanding Natural Beauty (AONB). The Countryside Commission has now been working for 50 years to conserve the beauty of the rural landscape, and encourage public enjoyment of the countryside.

It is one of the main government agencies with responsibility for conservation and recreation in the countryside. The 1949 Act gave the Countryside Commission two main aims:

- to enable wider enjoyment of the countryside by the public, especially by ramblers and cyclists
- to protect some of England and Wales' most beautiful scenery by the establishment of National Parks and AONBs.

This chapter examines three strategies used by the Countryside Commission to meet these two aims:
1 the creation of National Parks (pages 23–26)
2 the creation of National Trails (pages 27–28)
3 the development of forests as recreational areas (pages 28–31).

However, the Countryside Commission is just one of several organizations which make decisions about the use of the countryside. The various types of decision-makers are discussed in 'Who are the rural decision-makers?' below.

1 Form groups of two to three. On a sheet of A3 paper, consider each aim of the Countryside Commission, and the effects that each aim might have on rural communities. Present your findings to the class.

2 Do the two aims seem to be compatible? Explain your views.

Who are the rural decision-makers?

The management of a rural area is a complex issue. No one single person or group has control. Geographers are interested in the way in which resources such as education, health, housing or recreation are controlled and allocated by different individuals or agencies. There are at least three alternative models of resource control and allocation, and these may be applied to management of the rural resource:

- The managerialist model: Resources are controlled and allocated by managers or gatekeepers. These gatekeepers are professionals such as planners, accountants, housing officers and bank managers working in local government or in private commerce.
- The statist or structuralist model: The lives of local communities are influenced by decisions taken at national level by ministers in education, health, employment, etc. Decisions specifically on rural issues may be taken by the Department of Environment, Transport and the Regions (DETR). The landscape resource is allocated on a national scale by the Countryside Commission.

- The capitalist model: Most decisions are made and resources are allocated by private enterprise as part of the business economy. The privatization of elements of the welfare system (such as some aspects of the health service and even some prisons) proves that businesses, whether public or private, can be run and allocate almost any resource on a capitalist model.

There are numerous groups, at local and national level, who have an interest in rural management. They all have their part to play in influencing the decision-makers, or indeed allocating the landscape resource. Each of the three models may be used to explain the way in which these groups try to influence decisions and thereby allocate the landscape resource to the public. Figure 2.6 indicates how each model helps to explain the decision-making process in the rural environment.

▼ Figure 2.6 Decision-making in the rural environment.

Model	Type of decision-maker	Example of decision-maker in the rural environment	Type of influence exerted by this decision-maker
Managerialist	Professional manager	Highway authorities (the county council, metropolitan borough or unitary authority)	Responsible for recording the legal existence and location of public rights of way, and for ensuring that footpaths are open for use
Structuralist	Government agency	English Nature and the Countryside Council for Wales	Responsible for the National Nature Reserves: 88 000 hectares of land to which there is some public access
Capitalist	Commercial interest	Country Landowners' Association (CLA)	Most landowners are farmers who are actively involved in food production, management and conservation of the countryside. Private landowners are represented by the CLA. Landowners are obliged to maintain existing rights of way, but many resist further access because they wish to preserve the privacy of their land.

In considering the role of various bodies involved in rural management, it is useful to distinguish between government agencies such as English Nature and the Countryside Commission, and non-governmental organizations (NGOs). NGOs play an important role in the debate over public access to the countryside which may include:

- lobbying for change, for example the Ramblers' Association campaign to improve access to the countryside. An example of one of their adverts is shown in Figure 2.7. A further example is the Countryside Alliance, a collection of groups spearheaded by the British Field Sport Society (BFSS) who lobby the government to prevent anti-hunting legislation from being passed in parliament

- raising public awareness, producing educational materials and actively carrying out conservation in the field, for example the British Trust for Conservation Volunteers (BTCV)
- buying and/or managing land, for example the National Trust and the Woodland Trust who provide public access to large parts of their land.

Strategy 1: The creation of National Parks

The world's first National Parks were created in the USA in the late nineteenth century. These parks were wilderness regions largely unaffected by settlement and human activity. The National Parks of England and Wales, however, are not true wilderness regions. They have long histories of rural settlement and employment. However, they do contain some fine scenery which is protected by strict planning regulations to prevent the loss of landscape character by unnecessary or unsympathetic developments. National Parks created in countries such as India and Kenya have been established to protect ecosystems and wildlife from destruction. But the National Parks

Breathing Space.
(Don't take it for granted)

We don't. Help us keep Britain's breathing spaces open. Footpaths and coastline, high places, heaths and woodland. For walkers.

For over 60 years, THE RAMBLERS' lobbying and vigilance have been achieving wide-ranging rights of access to some of our most beautiful countryside.

Take a walk. Take a breather; consider the future. Invest in THE RAMBLERS.

The Ramblers
Working for walkers

Join us.
Your subscription brings our essential outdoor information Yearbook, the quarterly colour magazine, and membership of one of our 400 local groups.

Mr/Mrs/Miss/Ms

Address

Postcode   Date of birth

Tick box for membership type required
- Ordinary £18   Reduced* £9
- Family/joint £22.50 Joint reduced* £11.50
(for two adults at same address)
*Under 18/students/retired/disabled/unwaged

Donation £_____   I enclose £_____

- We occasionally exchange names (for use once only) with other organisations which may interest you. Tick if you prefer to be excluded.

Regd.Charity No.306089

1-5 Wandsworth Road, London SW8 2XX. Tel: 0171 339 8500

◀ **Figure 2.7** The Ramblers' Association advert which appeared in the national press in 1998.

◀ **Figure 2.8** The Lake District is England's most visited National Park.

of England and Wales do not have this conservation function. They were established for largely aesthetic reasons to do with landscape quality, not to protect or conserve wildlife.

There are eight National Parks in England and three in Wales. Each National Park is managed by a National Park Authority. The eight National Parks in England cover 7.6 per cent (9934 square kilometres) of the country. They offer exceptional opportunities for outdoor recreation. More than 61 million visitor days are spent in the National Parks each year. Figure 2.10 indicates the weight of visitor numbers. However, the sheer number of visitors can cause severe problems. Visitors usually arrive by car and create traffic congestion on narrow country roads. Walkers sometimes leave gates open, drop litter or allow their dogs to run free and worry livestock. Footpath erosion can scar the landscape. The conservationist, David Bellamy, has said of the Lake District that we are in danger of 'loving the countryside to death' (Figure 2.8). Visitor pressure on the landscape is described by using the concept of carrying capacity (see page 26).

The National Parks of England and Wales are not owned by the nation. In 1949 some had assumed that the government would purchase up to 10 per cent of the land in the National Parks. But this nationalization of land never happened. The majority of land in National Parks is in private ownership and much of it is inaccessible to the public (Figure 2.9). The 1949 National Parks and Access to the Countryside Act gave local planning authorities and landowners the means to provide public access to open country for open air recreation. Access may be by agreement or by order confirmed by ministers, but orders are very rarely made. Many landowners resist the creation of further public access to the countryside as they feel that their privacy should be respected. Some also think that visitors generally do not understand the countryside and that traffic and walking are both harmful to the environment.

▶ **Figure 2.9** Land ownership in the Peak District National Park (%).

Water companies	15
National Trust	12
Peak Board	4
Private ownership	69

National Park	Established (year)	Area (km²)	Population (1991)	Visitor days[1] (million/year, 1994)	Approved net expenditure (£m, 1997/8)
The Broads	1989	303	6 050	5.4	2.0
Dartmoor	1951	954	32 231	3.8	2.6
Exmoor	1954	693	12 160	1.4	2.1
Lake District	1951	2 292	43 180	13.9	3.8
The New Forest[2]	1990+	578	35 000	6.6	n/a
Northumberland	1956	1 049	4 040	1.4	1.6
North York Moors	1952	1 436	26 524	7.8	2.9
Peak District	1951	1 438	43 266	12.4	5.4
Yorkshire Dales	1954	1 769	19 220	8.3	2.8

▲ **Figure 2.10** Factfile: England's National Parks.

Note 1 Visitor numbers are difficult to estimate. Parks generally use different survey methods and different categories of visitor. The 1994 survey was the first to use similar methods in all parks. However, in most cases the figures are regarded as underestimates (except for The Broads, where the survey area was larger than the park itself).

Note 2 The New Forest has planning protection which recognizes its high landscape qualities. However, it has not been designated as a National Park so it does not have the equivalent administration or funding.

▼ **Figure 2.11** The National Parks of England.

Most of the Lake District's 2292 square kilometres consist of moorland and fell. Formed from glacial meltwater, the sixteen lakes, of which Windemere is the largest, are arranged like spokes of a wheel in the mountain valleys.

Exhibitions and events at Brockhole, the National Park Centre near Windemere, and a park-wide information centre network and events programme help visitors to understand and appreciate Britain's largest National Park. The 2896 kilometres of public rights of way provide unrivalled walking and climbing, from gentle lakeside strolls to testing mountain ascents.

Open heather moorland is the main feature of the North York Moors National Park. As the largest expanse of continuous heather moorland in England, it is home to precious wildlife such as Curlew and Merlin.

The eastern boundary of the Park is a 42-kilometre stretch of Heritage Coast, with high cliffs and wide sweeping bays.

Walking is the best way to get to know the Park, and there are more than 1609 kilometres of public footpaths and bridleways to choose from. For keen walkers, there is the Cleveland Way, a 174-kilomtere National Trail that loops from Helmsley to the coast. Information about the Park is available from three information centres – at Sutton Bank, Helmsley and the Moors Centre at Danby.

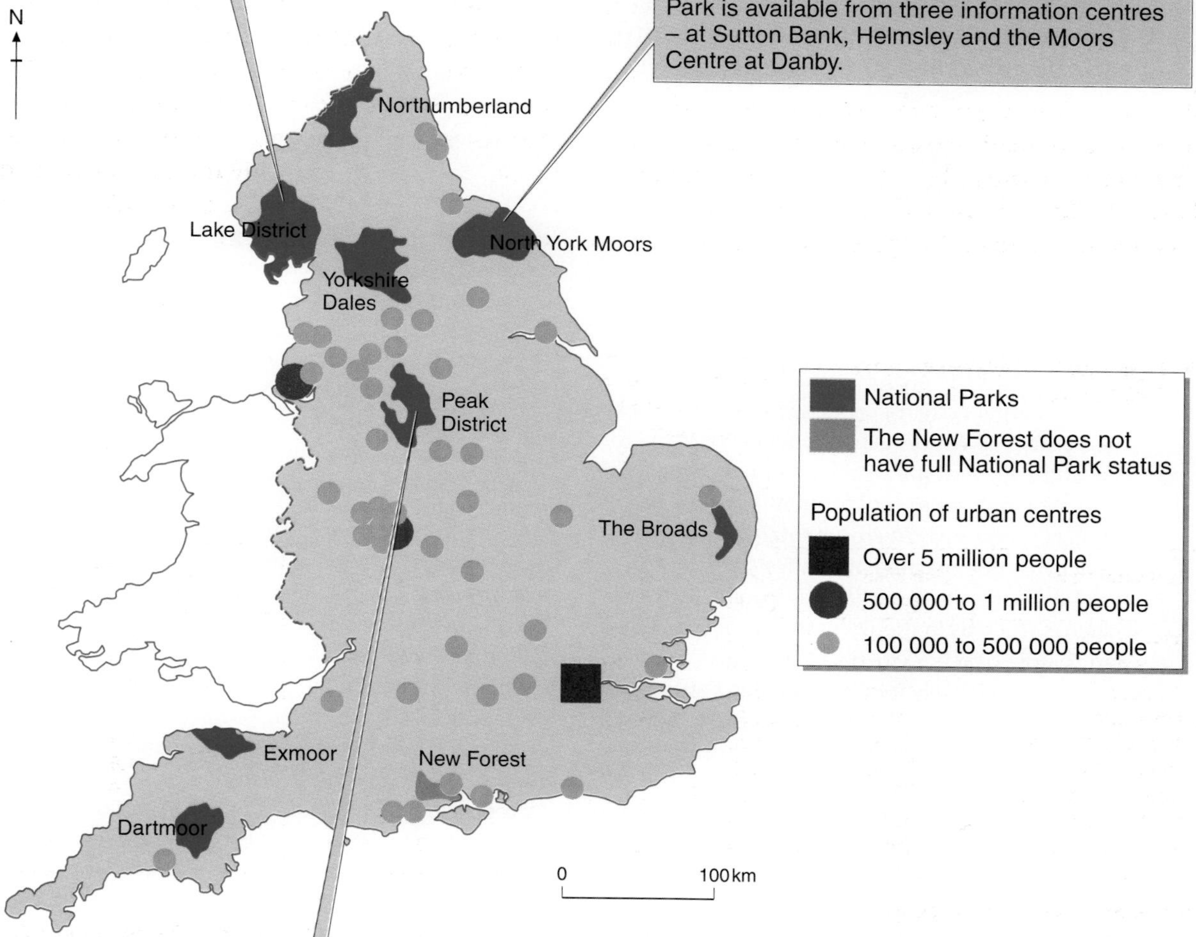

N

Northumberland

Lake District

North York Moors

Yorkshire Dales

Peak District

The Broads

National Parks

The New Forest does not have full National Park status

Population of urban centres

■ Over 5 million people

● 500 000 to 1 million people

● 100 000 to 500 000 people

Exmoor

New Forest

Dartmoor

0   100 km

Two distinct landscapes form the 1438 square kilometres of the Peak National Park. In the centre is the White Peak, with deep dales and undulating fields characteristic of limestone country. Around the north, east and west is the Dark Peak, a more sombre area of peat moorland, with edges of precipitous millstone grit, where heather and bracken predominate.

Access agreements cover more than 207 square kilometres of the northern and eastern moors. Along the crags of the Eastern Edges, and in the limestonedales, rock climbing is popular. Four old railway lines have been converted into attractive paths.

1 Make a large copy of the map in Figure 2.11. Use an atlas to identify maps of the UK to anwer these questions.
   a) Describe the distribution of England's National Parks.
   b) What landscape attributes do these parks contain? How far is each unique?

2 a) Add the major conurbations to your map.
   b) Which of the National Parks have significant population pressures and from where?

3 Consider the issues about access in the three parks described in Figure 2.11, such as ownership and visitor pressure. In groups research access issues in each of the remaining National Parks and present your findings to the rest of the class.

4 a) Why are the number of visitors to National Parks likely to be underestimated?
   b) What kinds of problems might this inaccuracy create for each of the National Park Authorities?

5 Suggest what the role of the National Park visitors' centres might be.

Carrying capacity

At what point do visitors to the countryside become a problem for local people or the environment? The concept of carrying capacity can be used to describe the point at which any site is placed under threat. There are three types of carrying capacity:

- Environmental carrying capacity is reached when the number of visitors causes physical damage to the landscape, habitat or ecosystem. Trampling of vegetation and erosion of footpaths are the usual physical symptoms. Environmental carrying capacity can be reached at relatively low visitor numbers if the site is particularly fragile. Salt marshes (such as in the Dyfi estuary in mid-Wales) and shingle beaches (such as Chesil Beach in Dorset) are relatively fragile and easily damaged by even a few visitors trampling across them. Forests, however, are more robust and can withstand greater visitor pressure.
- Psychological carrying capacity is reached when the pleasure and appreciation of visitors is reduced because of the number of people at the site. Open sites with wide horizons or long distance views are most easily affected. An example is Lulworth Cove in Dorset where enjoyment of the view can be spoilt by crowds of people or parked cars. Forests, on the other hand, have a greater psychological carrying capacity because people are hidden from each other's sight.
- Cultural and social carrying capacity is reached when the volume of visitors becomes a nuisance to local people. Excess noise, litter, traffic congestion and parking problems may all contribute to cause this nuisance. Visitors can sometimes be insensitive to local people, disregarding their privacy or trespassing on their land. It is this fear of trespassers that makes many country landowners resist greater access to the countryside.

Each type of carrying capacity has its own threshold levels which may be reached at different times. For example, local people may feel that visitors have become a nuisance at relatively low levels, well before visitors perceive each other to be a nuisance. The environment itself may not be damaged before an even larger number of people have visited the site.

Honey pot sites are those sites that attract most visitors. The huge volume of visitors at sites such as Dovedale in the Peak District or Lulworth Cove in Dorset can exceed all three types of carrying capacity. Resulting environmental problems at Dovedale have had to be carefully managed by the owners, the National Trust, (Figures 2.12 and 2.13). Problems are often seasonal, with the greatest pressures occurring over the spring and summer bank holiday weekends. However, managers of the countryside need to consider whether or not honey pot sites can be sacrificed for the greater good of other, more fragile sites.

▼ **Figure 2.12** Dovedale in the Peak District is a honey pot site visited by around 2 million people every year. On bank holiday weekends there are as many as 3000 visitors per hour.

▲ **Figure 2.13** The National Trust have improved the footpath using local limestone chippings. The 5 kilometres of path took ten years to lay and cost £250 000.

Strategy 2: National Trails

There are currently 169 000 kilometres of public footpaths, bridleways and byways in England. Some footpaths are hundreds of years old. Footpaths are maintained by landowners, but the local Highway Authority must ensure that they are open for use. The 1949 Act gave the Countryside Commission the power to create long-distance footpaths by joining existing rights of way into cross country trails. There are now ten National Trails in England (Figure 2.14), including the 412-kilometre Pennine Way. As can be seen in Figure 2.16, the trampling effect of walkers kills vegetation, exposing the soil to erosion and scarring the landscape.

▼ **Figure 2.14** National Trails in England (source: Countryside Commission website).

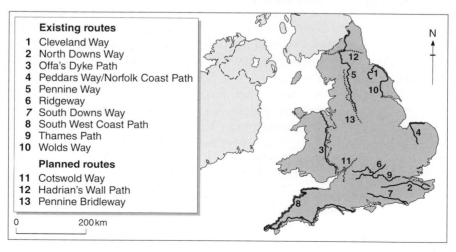

Existing routes
1 Cleveland Way
2 North Downs Way
3 Offa's Dyke Path
4 Peddars Way/Norfolk Coast Path
5 Pennine Way
6 Ridgeway
7 South Downs Way
8 South West Coast Path
9 Thames Path
10 Wolds Way

Planned routes
11 Cotswold Way
12 Hadrian's Wall Path
13 Pennine Bridleway

0   200 km

▼ **Figure 2.15** Factfile: England's National Trails (source: Countryside Commission website).

National Trail	Opened (year)	Length open to different users (km)			Total length (km)
		Horseriders	Cyclists	Motors	
Pennine Way	1965	83	83	38	412
Cleveland Way	1969	44	44	3	176
Offa's Dyke Path	1971	22	49	57	285
South Downs Way	1972	171	171	25	171
Ridgeway	1973	101	101	83	137
South West Coast Path	1973–78	–	35	45	962
North Downs Way	1978	122	122	74	246
Wolds Way	1982	44	44	20	130
Peddars Way/Norfolk Coast Path	1986	70	70	70	150
Thames Path	1996	14	16	1	344
Total					**3013**

▼ **Figure 2.16** An estimated 11 000 people walked this popular section of the Pennine Way in 1975. By 1985 this figure had more than doubled. The result is trampling of the slow growing moorland vegetation and footpath erosion.

1 a) Use the data in Figure 2.15 to draw a cumulative line graph of the total amount of footpath available to walkers between 1965 and 1996.
  b) What trends are shown on your graph?
  c) Have the Countryside Commission managed to satisfy rising demand for footpaths?

Strategy 3: Forests as recreational areas

Some large parcels of countryside are in public ownership. The Forestry Commission owns 395 600 hectares of forest. It has developed a policy of freedom of access through much of this forest. The Forestry Commission is a government agency which has the task of managing the UK's publicly owned forests. It was established in 1919 with two aims:

- to replant areas which had lost timber as a result of uncontrolled deforestation
- to make the UK more self-sufficient in timber.

Today, the UK has a forest cover of 7.5 per cent compared to a European average of 21 per cent. Forestry and timber processing are important rural employers, but UK forests produce only 8 per cent of our timber needs.

The 1968 Countryside Act recognized that forests have huge potential for recreation. Forests are a naturally robust habitat; their physical carrying capacity is high. Large numbers of people can visit a forest, but the trees conceal them from one another so there is a high psychological carrying capacity (see page 26). The 1968 Act gave the Forestry Commission wider responsibilities which included encouraging better access and recreational opportunities in its forests.

Much of the pioneering work on forest recreation was done in Grizedale forest in the Lake District. Over 90 per cent of the Forestry Commission's land is planted with conifers. Insensitive and monotonous planting schemes have traditionally angered conservationists and ramblers. But experiments in replanting small areas with deciduous trees have proved beneficial to both wildlife and recreation. The beauty of the landscape – its aesthetic value – can be enhanced by planting deciduous trees along the edge of the forest, and along forest walks. These strips of forest provide corridors along which insects, birds and mammals can move from one part of the forest to another. Mixed woodlands, therefore, have a greater biodiversity than single species forests (Figure 2.18). At Grizedale, the forest walks have been further enhanced by a long-term art project; a large number of sculptures have been created in the forest for walkers and cyclists to admire. An example is shown in Figure 2.17.

The timber industry is one which requires long-term investment. Investments made earlier this century in the purchase of land and planting of trees are only now making returns as the timber is harvested and processed. The Forestry Commission's assets are worth between £700 million and £1.4 billion. There has been much talk since 1990 about the possible privatization of the Forestry Commission. However, there is a fear that private industry would clear fell large areas of forest and make a quick profit on the long-term investments that are only now ready to make money for the national budget. It is feared that the privatization would also severely reduce the amount of forest available for recreation, since many private woodland owners have resisted public access agreements.

▲ **Figure 2.17** One of the many sculptures created by artists in Grizedale forest, Lake District.

▼ **Figure 2.18** Estimated numbers of breeding songbirds in two broad-leaved and two coniferous woods in Wales.

Bird	Broad-leaved wood (275 trees/ha)	Broad-leaved wood (375 trees/ha)	Coniferous wood (450 trees/ha)	Coniferous wood (275 trees/ha)
Chaffinch	7	4	4	7
Wren	13	15	34	9
Robin	7	4	3	3
Dunnock	–	–	2	–
Goldcrest	2	4	9	31
Coal tit	6	3	7	4
Blue tit	5	2	–	–
Great tit	6	2	–	–
Marsh tit	–	–	–	–
Nuthatch	3	–	–	–
Tree creeper	1	1	–	–
Blackbird	3	–	–	–
Song thrush	–	–	–	–
Mistle thrush	–	–	–	–
Wood warbler	3	3	–	3
Willow warbler	4	10	–	–
Blackcap	–	–	–	–
Chiffchaff	–	–	–	–
Pied flycatcher	6	7	–	–
Redstart	2	2	–	–

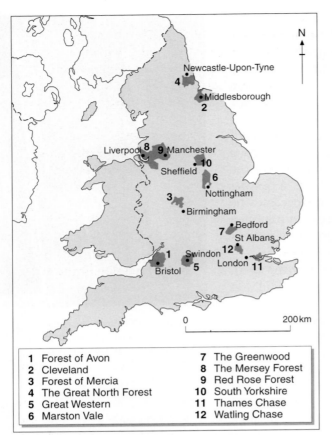

1 Forest of Avon
2 Cleveland
3 Forest of Mercia
4 The Great North Forest
5 Great Western
6 Marston Vale
7 The Greenwood
8 The Mersey Forest
9 Red Rose Forest
10 South Yorkshire
11 Thames Chase
12 Watling Chase

Community Forests

The Countryside Commission is now working in partnership with the Forestry Commission and local authorities to extend the UK's forest cover. It is hoped that the scheme will increase England's current 7.5 per cent tree cover to 15 per cent by the year 2050. In a long-term project, Community Forests are being planted on the outskirts of twelve major cities in England. The project started in 1990. Around 30 million trees have been planted already. Millions more will be planted in the twelve Community Forests shown in Figure 2.19. Most will be native, broad-leaved trees such as oak, beech and ash. One of the aims of the project is to create recreational areas within the new forests wherever possible.

◀ **Figure 2.19** Planned Community Forests in England (source: Countryside Commission website).

▼ **Figure 2.20** Factfile: Community Forests (source: Countryside Commission website).

Community Forest	Area (km²)	Number of local authority partners	Existing tree cover (%)	Population within 20km (millions)
Forest of Avon	573	6	5.9	1.0
Cleveland	255	5	6.9	1.0
Forest of Mercia (South Staffs)	210	5	6.4	4.0
The Great North Forest (south Tyne & Wear/ North East Durham)	160	5	8.0	1.0
Great Western (Swindon)	390	5	3.0	0.3
The Greenwood (Nottingham)	414	7	11.3	1.0
Marston Vale (Bedford)	158	3	3.6	0.5
The Mersey Forest	925	9	4.0	5.0
Red Rose Forest (Gtr Manchester West)	760	6	3.9	4.0
South Yorkshire	395	4	7.6	1.9
Thames Chase (East of London)	98	5	9.7	3.0
Watling Chase (South Herts/North London)	163	6	7.9	3.0
Total	**4501**	**66**	**–**	**25.7**

1 Form groups of three to four.
   a) Consider the arguments for and against privatizing the Forestry Commission.
   b) On a large sheet of paper identify the possible impacts of privatization on the economic, aesthetic, conservation and recreational functions of the forest.

2 Use Figure 2.18 to find:
   a) the total number of songbirds in each forest
   b) the number of species in each forest.
   c) How important are broad-leaved woodlands to songbirds?

3 Comment on the sampling method used in Figure 2.18. Why were four woodlands used in total?

4 Study Figures 2.19 and 2.20.
   a) 'Existing tree cover' and 'population within 20 kilometres' are two important criteria in selecting the location of a Community Forest. Justify the selection of each of these criteria.
   b) Is there a correlation between size of forest and population within 20km? Is there any reason why the two should be linked?
   c) Select and justify a location for another Community Forest.

5 Using the information in Figure 2.5 on page 20, evaluate the environmental and economic benefits of the Community Forest scheme.

Should the government extend the 'right to roam'?

One of the biggest demonstrations ever to take place in London was organized by the Countryside Alliance (see page 19). The demonstration highlighted several other rural issues including whether or not access should be improved for ramblers and others to enjoy the countryside. While the Country Landowners' Association (CLA) argue that there is already sufficient public access, the Countryside Commission has tried to balance the need for rural recreation against the environmental problems that visitors can create.

In 1998 the government began a consultation process on how best to extend public access to the countryside. The Ramblers' Association support the view that only new legislation will force landowners to create new rights of way across their land. However, the CLA insisted that a voluntary scheme was needed. The mistrust between the rural landowners and 'townies' is very much in evidence today, as Figures 2.21 and 2.22 show.

Landowners bar way to new 'right to roam'

Ian MacNicol, the CLA's president, said landowners wanted to meet public demand for new rights of way with new 'Permanent Paths', and that access would be managed to avoid conflict with wildlife, the environment and farming.

He said: 'The Government offers two options: we strongly support the voluntary approach and are totally opposed to the creation of a new statutory right of access. We believe that by using voluntary means, rather than compulsion, we can meet the changing demand for access in the way and in the places where it is actually wanted.'

But Kate Ashbrook, access chief of the Ramblers' Association, said: 'The CLA are going full speed back to the past: a past of failure, frustration and missed opportunities. But the Government is looking to the future now and has sought views on how all 4 million acres [1.6 million hectares] of mountain, moor, heath, down and common land can be opened up to walkers, subject to commonsense restrictions.'

Michael Meacher, the environment minister, who is a keen rambler, has warned farmers and landowners that if voluntary agreements do not work, the Government will introduce new laws forcing them to open up blocked footpaths.

Instead of the public being free to walk where they wish, the CLA's proposals would mean that local authorities would have to negotiate agreement with individual landowners. Mr MacNicol said that its proposals would cost the taxpayer between £4m and £7m, compared with more than £60m if a statutory right to roam was imposed.

▲ **Figure 2.21** From *The Independent* on Line, 3 June 1998.

Most landowners are hostile to any changes in access to the countryside, claiming that it will lead to waves of 'townies' wandering around trampling crops, disturbing livestock and leaving litter. Betty Bowes, regional secretary for Leicestershire, Rutland and Northamptonshire branch of the Country Landowners' Association (CLA) says that 'Right to Roam is effectively a nationalisation of landuse ... But it cannot just be a right for ramblers to roam. We don't want all and sundry roaming our land, especially not criminals, drug pushers and vandals.'

▲ **Figure 2.22** From *Guardian Education*, 17 February 1998.

1 Using Figures 2.21 and 2.22 and information from earlier parts of this chapter, outline the arguments:
   a) for extending the 'right to roam'
   b) for restricting any further access.

2 What are your views? Consider these:
   a) in groups of two or three
   b) as a whole class.

Ideas for further study

Investigate the roles of a variety of NGOs. Which NGOs are:

- involved primarily in lobbying the government
- producing educational materials
- actively conserving the rural environment
- in conflict with the views of other government or non-government organizations?

You will need to visit their websites, or write to them asking for information about their work.

Summary

- The process of counterurbanization has resulted in a number of contentious issues in the UK rural environment. Among these is the issue surrounding the 'right to roam'.
- Increased recreation in the rural environment brings some benefits to the rural economy. However, associated environmental problems, such as traffic congestion and footpath erosion, grow worse as visitor numbers increase.
- Decisions about this, and other rural issues, are taken by a variety of government agencies and NGOs. Some groups take on lobbying and educational roles, while other agencies have legislative powers or conduct practical conservation techniques.

References and further reading

Rural recreation and the changing rural landscape are discussed in:

John Chaffey (1994) *A new view of Britain*. Hodder and Stoughton.
Robert Prosser (1994) *Leisure, recreation and tourism*. Collins Educational. (especially chapters 7 and 8)
Susan Owens and Peter L Owens (1994) *Environment, resources and conservation*. Cambridge University Press. (especially chapter 6)

The Countryside Commission has a useful website that describes the role of the Commission in relation to National Parks, National Trails, Community Forests, etc. at http://www.countryside.gov.uk

A number of groups interested in rural issues also have websites. These include The Ramblers Association at http://www.ramblers.org.uk and the Country Landowners Association at http://www.cla.org.uk
A number of other organizations can all be contacted via an environmental website called Greenchannel. This site will link you to organizations such as the British Trust for Conservation Volunteers (BTCV) or the Council for the Protection of Rural England (CPRE) and numerous others. Greenchannel is found at http://www.greenchannel.com

Managing the rural environment in LEDCs

Introduction

We have seen in Chapters 1 and 2 that counterurbanization is creating conflicts over the use of rural space. Shrinking cities and growing rural populations are causing management problems in rural environments in the UK and other more economically developed countries (MEDCs). However, in the poorer countries of Africa and Asia the rural management issues are quite different because the population dynamics are different. Rural to urban migration takes place on a massive scale in many less economically developed countries (LEDCs). This rapid population growth in the cities puts pressure on work, sanitation and housing. But what effect does migration have on the rural areas that the migrants leave behind? Does the loss of so many people cause economic decline? Or are the migrants simply replaced by the natural increase of the population? Whichever is the case, how is the rural environment coping with the demands of a changing population?

This chapter examines the environmental issues facing rural regions of Kenya and India. Case studies on a regional scale show how population change has affected three vital rural resources: woodfuel, soil and water. Kenya and India have been chosen because they both have predominantly rural populations (their urban populations will be 32 per cent and 34 per cent respectively by the year 2000) and rapid rural to urban migration. These case studies also highlight the role played by NGOs, many of whom are using self-help schemes and appropriate technology to try to solve rural problems.

Patterns of rural and urban population change

Urbanization is occurring most rapidly in the world's poorest LEDCs. In most parts of Africa and South Asia, the population is still predominantly rural. However, rural to urban migration and natural increase in the population are combining to create rapid urban growth. Thus the African urban population increased from 19 per cent in 1960 to 33 per cent in 1990 and is expected to rise to 39 per cent by the year 2000 (Figure 3.1). The most rapid rates of urban growth occurred during the 1980s. The rate is beginning to decrease as natural increase begins to slow in both urban and rural areas.

▼ **Figure 3.1** Estimated urbanization, urban and rural population in Africa, 1960–2000 (source: Binns, 1995 p. 49).

	Total population				Urban population			Rural population	
Date	Millions	Interim annual growth rate (%)	Level of urbanisation (%)		Millions	Interim annual growth rate (%)		Millions	Interim annual growth rate (%)
1960	281		19		53			228	
		2.4				4.2			2.0
1965	316		21		65			251	
		2.6				4.5			2.1
1970	359		23		81			278	
		2.9				4.7			2.3
1975	414		25		102			312	
		3.0				4.8			2.3
1980	479		27		129			350	
		3.0				5.0			2.2
1985	555		30		165			390	
		3.0				4.9			2.1
1990	642		33		209			433	
		3.1				4.9			2.1
1995 projected	747		36		266			481	
		3.0				4.8			1.9
2000 projected	866		39		337			529	

How might population growth affect the rural area?

The population changes described in Figure 3.1 are averages for the whole of Africa. They disguise regional variations, but they do indicate that, despite the effects of rural to urban migration, rural populations continue to grow. Population pressure in rural areas is thought to contribute to socio-economic problems such as under-employment and seasonal unemployment. Environmental problems such as soil degradation through overcollection of firewood, overgrazing or overcropping are also thought to be related to rural population increase. Some of the pressures on the rural environment are summarized in Figure 3.3.

◀ **Figure 3.2** Severe gully erosion occurs where rural land is badly managed. Does population pressure on the rural environment automatically lead to this kind of problem?

▼ **Figure 3.3** Pressures on the rural environment.

Population growth	Political pressures	Urban pressures
• Overcropping, overgrazing and reduction of soil fertility • Poverty and poor health • Shortage of fuelwood • Soil erosion and gullying	• Demands to grow cash crops for export • Political instability • Civil war • Planting of land mines on agricultural land	• Attitudes of an urban elite who ignore rural issues • Demand for rural resources such as food, timber and water • Urban sprawl consumes agricultural land • Prevention of traditional nomadic migration

The effect of rural to urban migration

Rural to urban migration has reduced rural population pressure on land and the environment. It is assumed that social and environmental problems might be reduced as population pressures decline. However, the picture is not clear. Many argue that people are themselves an important resource. People work together to find new, more appropriate methods of solving rural problems. This optimistic view of the effect of population growth is credited to Boserup (see page 42).

Circular migration is common in Africa, especially in the Sahel and savanna climate zones. Rural employment is seasonal so workers migrate to urban areas during the dry season. Migrants earn money which can be sent to the rural family and invested in the farm. Circular migrants also reduce demand on village food and water stocks during the long dry season. A detailed study of the Machakos district of central Kenya has shown that circular migration has had enormous benefits for the rural area. Money sent home by migrants to their families has been invested in improving terraces and in tree planting schemes. It has also been used to diversify farm products. The growing cities of Mombasa and Nairobi have created extra demands for agricultural products such as meat, fruit, vegetables, timber and charcoal which has meant increased incomes for farmers in Machakos. A more detailed description of this study can be found on page 40.

How are rural and urban regions linked?

We have seen that rural and urban regions of LEDCs are linked by circular and permanent flows of migrants. However, there are many other urban–rural links. Some of these will benefit the rural area, while others degrade the rural environment (Figure 3.4).

▼ **Figure 3.4** Summary of urban–rural links.

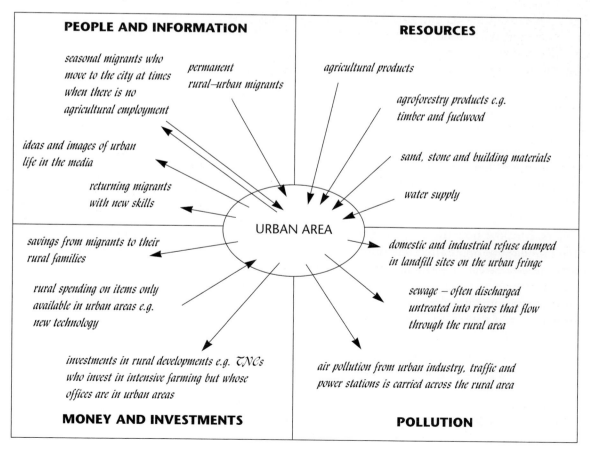

PEOPLE AND INFORMATION

seasonal migrants who move to the city at times when there is no agricultural employment

permanent rural–urban migrants

ideas and images of urban life in the media

returning migrants with new skills

URBAN AREA

savings from migrants to their rural families

rural spending on items only available in urban areas e.g. new technology

investments in rural developments e.g. TNCs who invest in intensive farming but whose offices are in urban areas

MONEY AND INVESTMENTS

RESOURCES

agricultural products

agroforestry products e.g. timber and fuelwood

sand, stone and building materials

water supply

domestic and industrial refuse dumped in landfill sites on the urban fringe

sewage – often discharged untreated into rivers that flow through the rural area

air pollution from urban industry, traffic and power stations is carried across the rural area

POLLUTION

What are the environmental consequences of rural to urban migration?

- Small-scale farmers are put under pressure to increase production of cereals or market garden produce. This leads to overcultivation, overgrazing and overabstraction of water for irrigation.
- Woodfuel is used in both rural and urban areas. Farmers usually produce their own timber or trade with their neighbours. But the demand for woodfuel in urban areas has lead to a commercialization of this subsistence crop. Wood is now an important commodity, grown in rural areas and consumed in cities. The urban poor suffer as the cost of woodfuel accounts for as much as 30 per cent of their income. Rural areas suffer as forests are depleted; woodfuel is supplied to cities such as Kano (Nigeria), Khartoum (Sudan) and Nairobi (Kenya) from a radius of up to 400 kilometres.
- Many cities rely on the rural production of charcoal. This also depletes rural forests and hedgerows. Charcoal is lighter and therefore cheaper to transport than wood. But it is 40 per cent less efficient than fuelwood.
- Urban expansion often consumes valuable, productive farmland on the urban fringe. In a few cases, the loss of land has severe consequences for food production. Only about 3 per cent of Egypt is cultivable: a narrow strip of land along the Nile and in the Nile's delta. The ancient Egyptians knew the importance of the 'greenland' for food production, and built settlements on marginal or desert land. Much of Cairo's recent rapid growth has been at the expense of the greenland. Despite strict planning control, farmers sell land for a quick profit to developers for housing. Between 1950 and 1984 an estimated 12 per cent of Egypt's farmland was lost to suburban sprawl.
- Urban waste is tipped in rural landfill sites polluting both air and water. During the dry season dust particles from open tips contaminate the atmosphere. During the wet season water percolating through the tip leaches out harmful waste products to contaminate the ground water. Rubbish from Dakar, Senegal's capital, is dumped in an unmanaged rural landfill site situated between two freshwater lakes. Effluents from the tip are contaminating groundwater which is used for drinking and for irrigating market garden crops.

---

1 Study figure 3.1. What is happening to the absolute number of people living in rural Africa?

2 a) In pairs, consider how the growth of the rural population might affect the rural environment. Select one African country and research:
   - its current population trend
   - the impact of this trend on rural areas.

b) As a class, discuss whether current population trends threaten or benefit the rural environment in Africa.

3 What is circular migration and how might it benefit rural areas? Consider its effects on your chosen country.

4 Which rural–urban links seem to affect the rural areas adversely?

Rural management in the Kenyan highlands

The Kenyan highlands have a cooler and wetter climate than other parts of this East African country (Figure 3.6). The rainfall follows a bimodal regime: there are two rainy seasons, one in March to May, the other in October to December. The volcanic soils are generally fertile. As a consequence, the highlands have the most productive agricultural land and the highest population densities in the country. Rural population densities are as high as 1000 persons per square kilometre in some districts. Despite rural to urban migration, rural populations continue to grow rapidly and so average farm sizes are declining.

The demands of the rural population on land, timber and water resources have presented Kenyans with serious rural management issues. For example, the deforestation of natural closed forest has been rapid. The amount of Kenya that is forested has fallen from 20 per cent to around 4 per cent in 50 years. Timber for building poles and fuelwood has become more scarce, and the soil is beginning to lose its fertility and become more susceptible to erosion as a consequence. The two case studies which follow will demonstrate how these rural issues are being managed.

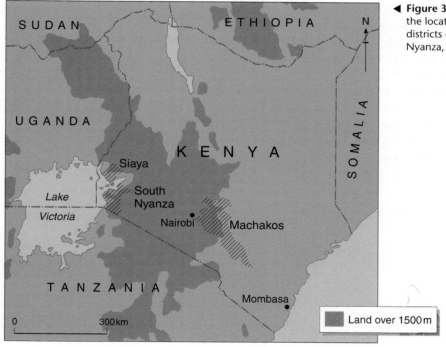

◀ **Figure 3.5** Kenya's relief and the location of the case study districts – Siaya and South Nyanza, and Machakos.

◀ **Figure 3.6** The climate of Nairobi is typical of the Kenyan highlands.

Average daily temp (°C)			
	Max	**Min**	**Precipitation (mm)**
Jan	25	12	38
Feb	26	13	64
Mar	25	14	125
Apr	24	14	125
May	22	13	158
Jun	21	12	46
Jul	21	11	15
Aug	21	11	23
Sep	24	11	31
Oct	24	13	53
Nov	24	13	109
Dec	23	13	86

Case study 1: Agroforestry in the highlands of Siaya and South Nyanza districts

The south-west region of Kenya is the most densely populated part of the country. This mountainous area has enough rainfall for farmers to grow a variety of crops such as sugar cane, cotton, maize, coffee and tea.

Up until the mid-twentieth century farmers traditionally collected forest products from surrounding woodlands such as timber poles for building or fuelwood. Hedges were planted to mark boundaries and provide shade, but trees were rarely planted on farms. As the population of the region grew, more and more forest was cut and access to natural forest products declined. The colonial government (1902–63) actually restricted the planting of trees on cropland through a system of fines and restrictions to loans and credit. It is only since independence in 1963 that a more widespread and intensive use of agroforestry has meant that the number of trees grown on farms is beginning to increase.

Agroforestry is the farm production of forest products. The trees provide valuable products such as fruit, building poles and fuelwood. The trees are planted in dense hedgerows, in between crops such as traditional fodder, vegetable or cash crops, or in plantations on the farms. Agroforestry has helped to reverse the impact of deforestation: while the percentage of tree cover has declined, tree density on farms has more than doubled in recent years.

Agroforestry has a number of environmental and economic advantages to the farmer:

- Products, such as timber poles for building, can be used by the farmer (therefore reducing household expenses), or sold to provide income for the farm.
- Trees planted in rows provide shade for crops and a wind break, so modifying the microclimate.
- A wide biodiversity is created. Different trees attract different insects which are valuable for the pollination of crops. Some trees attract insects that act as a biological pest control for crops.
- Soil management is improved. Leaves and clippings are used as a mulch to improve the soil. Trees can help with:
  - retention of soil moisture
  - soil stabilization on slopes
  - nitrogen fixing in soil
  - nutrient cycling.

The Agroforestry Extension Project

The Kenyan government and an NGO, CARE International, sponsor and promote the Agroforestry Extension Project in the highlands of Siaya and South Nyanza districts (Figure 3.5). The project began in 1984. A detailed study (of 336 households) showed that the project had achieved significant results in just five years. The average number of free standing trees on each farm had risen from 227 to 513. The average length of dense hedges had increased from 245 metres to 317 metres.

The diversity of trees planted on the farms was great: 170 different species were recorded in the survey, of which 39 per cent were traditional species indigenous to the area. Figure 3.7 gives details of plantings for selected species. Farmers have evidently chosen to plant a variety of trees, choosing different species to perform different functions. For example, Markhamia and Leucaena are planted in lines between crops. These trees are regularly pruned to control shade and provide green manure for the crops. A wide variety of fruit trees, including citrus, are grown in small numbers on each farm. They supplement the household diet, especially food for children, and provide fruit that can be sold in the market for cash. Eucalyptus is also grown as a cash crop (Figure 3.9). This fast growing tree quickly produces timber and poles that can be sold in urban construction markets. The study indicates that farmers on different incomes use trees for different purposes. For example, the poorest farmers are most likely to grow trees for woodfuel, while those on average incomes are more likely to grow trees for their fruit or timber. The five most commonly sold forest products are indicated in Figure 3.8.

Tree species	Trees established before the project began		Trees established during the first five years of the project		Total number of trees in 1989	
	Number of trees	Percentage	Number of trees	Percentage	Number of trees	Percentage
Agave[1]	6 096	8	1 760	2	7 856	5
Euphorbia[1]	5 065	7	1 676	2	6 741	4
Lantana[1]	5 252	7	636	1	5 888	3
Acacia[1]	2 843	4	90	·1	2 936	2
Psidium[1]	2 659	3	1 505	2	4 265	2
Markhamia[2]	10 747	14	12 417	13	23 164	13
Cassia[2]	2 282	3	5 798	6	8 707	5
Citrus[2]	670	1	2 461	3	3 131	2
Eucalyptus[2]	12 698	17	20 771	22	33 469	19
Luecaena[3]	348	·1	16 638	17	16 986	10
Terminalia[3]	2 915	4	2 731	3	5 844	3

[1] species native to Kenya
[2] species introduced between 1900 and 1950
[3] species introduced since 1970

▲ **Figure 3.7** The impact of the Agroforestry Extension Project on numbers of selected trees (source: S J Scherr, 'Agroforestry', *National Geographic Research and Exploration,* 10 (2) (1994) pp. 145–57).

▶ **Figure 3.8** Percentage of families who are selling forest products (source: S K Scherr, 'Agroforestry', *National Geographic Research and Exploration,* 10 (2) (1994) pp. 144–57).

Building poles	51
Fruit	47
Fuelwood	32
Tree seedlings	22
Tree herbs	10

◀ **Figure 3.9** Eucalyptus (gum) trees are native to Australia. They have been planted in many African countries because they are fast growing. However, they also require more water than native trees and are invasive and aggressive in their use of soil moisture when competing against native species of tree.

▼ **Figure 3.10** Use of trees in the highlands of Siaya and South Nyanza districts (%) (source: S L Scherr, 'Agroforestry', *National Geographic Research and Exploration,* 10 (2) (1994) pp. 144–57).

	Trees established before the project	Trees established during the project
Boundary markers	5	<1
Building poles	35	38
Charcoal	<1	<1
Fodder	1	1
Fruit	7	8
Fuelwood	24	10
Green manure	<1	15
Live fencing	10	9
Shade	4	4
Timber	8	9
Windbreaks	2	1
Other	3	3

1 How important does agroforestry seem to be to:
   a) the local environment
   b) the local economy.

2 Study Figure 3.7 carefully.
   a) What appears to be happening to the balance between native and non-native trees?
   b) Draw a graph to show this change.

   c) Suggest how the introduction of non-native trees might have both positive and negative effects on the environment.
   d) What other strategies might be worth attempting?

3 Using Figure 3.10 describe the main uses for agroforestry trees. How are these uses changing? How might this affect the balance between native and non-native trees?

Case study 2: Soil conservation in Machakos district

Machakos is a district to the east of Nairobi (Figure 3.5). It has been the subject of an extensive study into the effects that population increase have on the rural environment. The study used historical records dating back to the 1930s to paint a picture of change which covers the last 70 years of the twentieth century. It evaluates the role played by various government organizations and NGOs in the application of soil conservation techniques. The study reaches the conclusion that environmental improvements have been achieved while rural populations have grown and farming incomes have increased. How has this been achieved?

In order to appreciate this case study fully, you need to be aware of Kenya's colonial history. You also need to understand the distinction between those who take an optimistic view of population growth, and those who are pessimistic (see page 42).

Kenya's colonial history

Kenya has an extremely long and important pre-colonial history. The earliest remains of human ancestors in the world were found on the shores of Lake Turkana. Kenya's recent history has had a deep impact on its rural environment. Kenya's colonial past is one factor which has influenced rural population pressure. It also had an affect on both soil conservation techniques and methods of agroforestry. A summary of Kenya's colonial and post-colonial history is given in Figure 3.11.

▶ **Figure 3.11** Kenya's colonial and post-colonial history.

1887	Sultan of Zanzibar leased part of Kenya's coast to the British
1896	British Foreign Office assumed direct control of most of Kenya
1901	Railroad was completed from Mombasa to Lake Victoria
1902	All Kenya became a dependency under the control of the Foreign Office
1920s–50s	Main period of settlement by European farmers in the Kenyan highlands
1938–54	Only European farmers were allowed to grow coffee
1952–56	Mau Mau armed resistance to colonial rule
1963	Independence
1964	Republic declared; Jomo Kenyatta became the first president
1978	Kenyatta died; Daniel Arap Moi became the new president

Population densities increased particularly rapidly during the period of British colonial rule. This was due partly to natural increase, but also to government redistribution of land. The British rulers of Kenya restricted African access to both the best quality agricultural land and areas of grazing. European farmers were allowed to settle in the Kenyan highlands where conditions were best for agriculture. This was the Scheduled Areas (White Highlands) policy and as a result, 440 000 hectares of the most productive land was acquired by British and South African farmers and transnational agri-businesses. The British guaranteed land for tribal use. In effect this meant that Kenya's tribes were restricted to certain areas known as reserves. However, the British decreed that 'unoccupied' land should become Crown Land. The 'unoccupied' lands were the semi-arid and arid savannas of north and eastern Kenya. The indigenous nomadic people now had to apply for a permit to graze their cattle on these Crown Lands. The real function of the Crown Lands was to provide an area for Europeans to hunt big game. Years later much of this land became Kenya's game reserves and now attracts tourists with cameras instead of guns.

Squeezed out of both the highlands and the dry plains, the African tribes found themselves restricted. Population densities increased and farmland was put under greater pressure. Periods of fallow had to be abandoned and the land had to be used more and more intensively. Traditional farming methods and crops were replaced by the intense production of cash crops. Farms became smaller as children inherited land from their parents and plots of land became increasingly subdivided. The largest tribal group in Kenya, the Kikuyu, organized armed resistance to colonial rule. During the four years of this Mau Mau rebellion 11 000 Kikuyu were killed and 80 000 imprisoned in labour camps. But it was largely as a result of their protests that land was eventually redistributed to African people and Kenya gained its independence.

Population growth: Malthus and Boserup

Growth rates of populations, of both countries and cities, are obviously of great interest to planners and environmental scientists. Rapid growth rates produce steep exponential curves as can be seen in Figure 3.13. Exponential growth means that the population increases more and more rapidly and produces short doubling times of the population. Many social scientists believe that exponential growth results in severe resource shortages and pollution problems.

Thomas Malthus, writing in the 1790s, was extremely pessimistic, believing that rapid population growth would lead to severe problems of food supply with consequent malnutrition and famine (Figure 3.13). At the time he was writing the population of England and its cities was growing so rapidly that the population of Manchester doubled in the 28 years between 1773 and 1801 (Figure 3.12). Cities at that time relied more closely on food production in the neighbouring rural area than they do today, as food is now imported from all parts of the world. Malthus predicted that exponential growth of the world population would outstrip food supply. The result would be malnutrition, epidemics, famine or even war.

An environmental debate about the growth of population and the consumption of limited resources was stimulated by the Apollo flights to the moon in the 1960s. Images of the Earth in space made environmentalists aware of the precious nature of the Earth.

1773	36 250
1801	72 250
1851	303 500
1901	607 000

◀ **Figure 3.12** Exponential population growth in Manchester at the beginning of the nineteenth century.

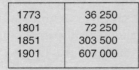

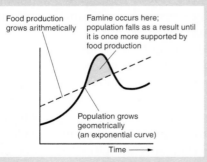

◀ **Figure 3.13** The prediction of Thomas Malthus.

Food production grows arithmetically

Famine occurs here; population falls as a result until it is once more supported by food production

Population grows geometrically (an exponential curve)

Time ⟶

Malthusian views were explored again in several books including the influential *The Population Bomb* (1968) by Paul Ehrlich (New York, Ballantine Books). These Neo-Malthusian views influenced the thinking of some governments and organizations during the 1970s and 1980s. The Indian and Chinese governments, for example, attempted to control population growth using strict ante-natalist policies. The neo-Malthusian view is that uncontrolled population growth is the root cause of resource and environmental problems. They argue that rapid increase in population puts a strain on services and infrastructure, such as food supply, energy, housing, sewage systems and transport networks. Where these stresses occur within the urban system, they may then lead to further problems, such as poor health and increased violence. These views tend to influence the press and general public leading to gloomy predictions about the fate of the world's mega-cities or of the world environment as a whole.

Other social scientists take an alternative, much more optimistic view of the consequences of rapid population growth. Such a view was put forward by Esther Boserup, a Danish economist in another influential book written in 1965. Boserup argued that improved agricultural techniques such as soil conservation and the use of higher yeilding varieties of grain allow farmers to greatly improve food output as population rises. She, and others, believe that resource shortages force people to invent new processes and find new technologies to increase food production, create cheaper and more efficient energy and transport systems, and overcome resource and pollution problems. People are innovators and so are themselves a valuable resource. This idea has encouraged some governments (such as that of Singapore) to adopt pro-natalist policies, in which people are encouraged to have larger, not smaller, families. The 'Boserupian' view is that people and their innovations therefore lead to improved standards of living. This contrasts with the pessimistic view of the neo-Malthusians in which society slumps towards famine and poverty.

Machakos: the physical characteristics of the district

The climate in Machakos is seasonal with two periods of rain which provide two growing seasons: March–May and October–December. Mean annual rainfall is in the range of 500–1000 millimetres depending on relief. The upland areas have more rain than the lower plateau that lies to the east and south of the district. The rainy seasons are unpredictable and several droughts have been recorded.

▶ **Figure 3.14** Agricultural areas in Machakos (source: Binns, 1995, Fig. 7.2).

Area A: 9% of the district; best agricultural area in the district; suitable for coffee

Area B: 40% of the district; intermediate agricultural area; suitable for cotton, agroforestry and fruit growing

Area C: 50% of the district; poorest agricultural area; suitable for livestock and growing maize and beans

▼ **Figure 3.15** Relief of Machakos district (source: WJEC A level G5 paper, 1998).

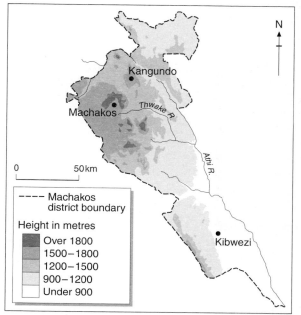

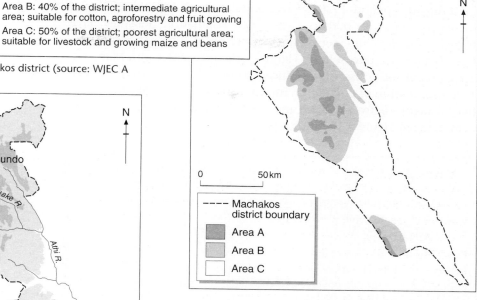

Machakos district boundary
Area A
Area B
Area C

▼ **Figure 3.16** Mean annual rainfall of Machakos district (source: WJEC A level G5 paper, 1998).

Long rainy season March–May

Short rainy season Oct–Dec

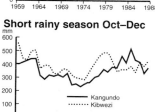

◀ **Figure 3.17** Rainfall in Kangundo and Kibwezi, 1959–88 (source: WJEC A level G5 paper, 1998).

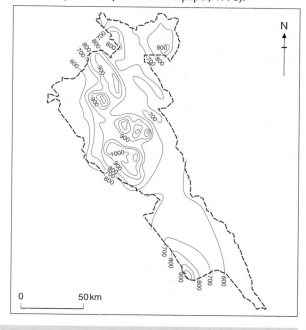

1 Use the three maps (Figures 3.14, 3.15 and 3.16) to produce an annotated sketch map of Machakos district. Identify the three agricultural areas on your map. Label these with a description of the relief and rainfall in each area.

2 Study Figure 3.17.

a) What pattern does the rainfall follow?

b) Does one area have more reliable rainfall than the others? Justify your answer.

3 Suggest reasons for the attractiveness of this part of Kenya to European settlers in the nineteenth century.

Population increase in Machakos

During the period 1930–90 the population of Machakos district increased from 238 000 to 1 393 000. Population densities have increased to such an extent that each person in the district now has less than one hectare of land (Figures 3.19 and 3.20). Colin Mahler, a government soil conservation officer, visited the district in the 1930s. The combination of population increase and poor rainfall (there were five droughts between 1929 and 1939) lead Mahler to make the comments in Figure 3.18 in his report of 1937.

> The Machakos Reserve is an appalling example of a large area of land which has been subjected to an uncoordinated and practically uncontrolled development by natives whose multiplication and the increase of whose stock has been permitted, free from the checks of war and largely from those of disease, under benevolent British rule.
>
> Every phase of misuse of land is vividly and poignantly displayed in this Reserve, the inhabitants of which are rapidly drifting to a state of hopeless and miserable poverty and their land to a parching desert of rocks, stones and sand.

◀ **Figure 3.18** From Mahler, 'Soil erosion and land utilisation in the Ukamba Reserve (Machakos)', report to the Dept of Agriculture, Mss Afr. S.755 Rhodes House Library, Oxford (1937), quoted in Binns, 1995.

During the 1930s, 1940s and 1950s European farmers grew coffee in the Machakos highlands where the climate was best for farming. The indigenous Akamba people were restricted to a densely populated reserve by the White Highlands policy. Furthermore, they were not allowed to settle on the plateau to the south and east which had been designated as Crown Lands. After independence in 1963, the Akamba settled in the Crown Lands. Some took over the management of former British coffee farms. The amount of space available for settlement in Machakos doubled.

▼ **Figure 3.19** Population growth of Machakos district, 1932–89 (source: Tiffen *et al.*, *More People, Less Erosion. Environmental Recovery in Kenya*, Wiley (1994).

Year	Total population (000s)	Growth rate (% per annum)	Average density (ha/person)
1932	240	–	2.66
1948	366	2.68	1.93
1962	566	3.17	1.38
1969	707	3.22	1.92
1979	1023	3.76	1.33
1989	1393	3.09	0.97

1 Study the material on this page and in Figure 3.20. Compare it with Malthus' and Boserup's views (page 42). Whose predictions seem to be coming true?

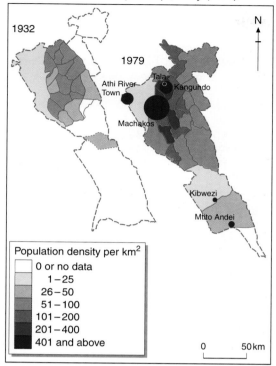

▼ **Figure 3.20** Population density of Machakos district in 1932 and 1979 (source: Tiffen *et al.*, *More People, Less Erosion. Environmental Recovery in Kenya,* Wiley (1994).

1 a) Use Figure 3.19 to draw a line graph of population increase in Machakos.
  b) How long did it take for the population to double after 1932, and then to double again?

2 a) Use Figure 3.19 to draw a single graph showing both growth rate and average population density in Machakos.
  b) Label the point at which the Akamba were given back their land. Explain the significance of this point on the graph.

3 Study the language used by Mahler in his 1937 report (Figure 3.18). In what ways is it:
  a) neo-Malthusian (see page 42)?
  b) Eurocentric?

4 Were Mathus' predictions already coming true in 1937?

Conserving the soil resource

Soil conservation methods have been used in Machakos since the 1930s (Figure 3.22). Two main techniques of terracing have been used (Figure 3.21):

- narrow based terraces in which soil is thrown downhill from a ditch
- wider bench terraces in which soil is thrown uphill – a technique known as *fanya juu* (which means throwing upwards).

▼ **Figure 3.21** The two main types of terracing used in Machakos.

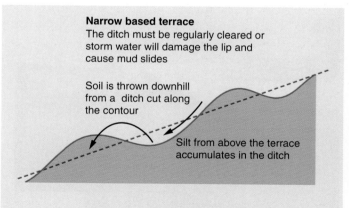

Narrow based terrace
The ditch must be regularly cleared or storm water will damage the lip and cause mud slides

Soil is thrown downhill from a ditch cut along the contour

Silt from above the terrace accumulates in the ditch

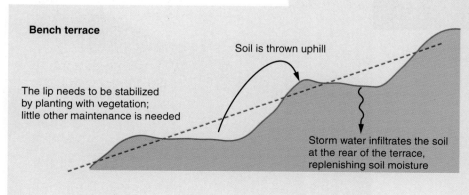

Bench terrace

Soil is thrown uphill

The lip needs to be stabilized by planting with vegetation; little other maintenance is needed

Storm water infiltrates the soil at the rear of the terrace, replenishing soil moisture

1937	Soil conservation service established. Early experiments were with lines of trash (cut vegetation) but it was thought to encourage pests and diseases.
1937–40	Bench terraces were constructed by compulsory labour groups. One person per household had to labour on soil conservation two mornings per week. Most labourers were women, since many men were urban migrants.
1940–55	Narrow based terraces were constructed by the labour gangs.
1946–62	African Land Development Board spent £1.5m on land improvement in Machakos. Much was spent on distribution of hand tools for building terraces.
1948	The compulsory labour gangs were becoming increasingly unpopular. In Murang'a district 2500 women refused to take part in terrace construction. They were arrested and fined.
1956	Compulsory labour gangs were replaced by voluntary community work parties (known as *mwethya*). Neighbours helped each other to construct terraces of their own choice. Most chose to construct bench terraces. Women often took leadership positions within *mwethya* groups.
1959–65	Terracing became unpopular. Terrace construction was associated with the colonial era. Many narrow terraces were allowed to decay.
1965–78	Terrace construction using *mwethya* self-help groups became widespread again. Aerial photos taken in 1978 show almost 100 per cent terracing in many long-settled areas (agricultural area B).
1978–98	Various NGOs began to involve themselves in soil conservation in response to severe droughts in the early 1970s. The Swedish International Development Agency (SIDA) began to provide hand tools, food aid and soil conservation experience in 1978 and is still working in Machakos alongside the Kenyan government. The European Community donated aid until 1988. Aid agencies contributed to the construction of 4500 kilometres of terraces per year throughout the 1980s. However, the technique had become so widely accepted that a further 4000 kilometres per year were built by farmers without any assistance from NGO aid. Community involvement is essential if these kinds of scheme are to work (Figure 3.27).

▲ **Figure 3.22** Timeline of soil conservation techniques.

▲ ▶ **Figure 3.23** A degraded hillside photographed in 1937 has been improved by terracing and agroforestry techniques.

▶ **Figure 3.24** Land use change in Masii (in agricultural area B) between 1948 and 1978 (source: Tiffen *et al.*, *More People, Less Erosion. Environmental Recovery in Kenya*, Wiley (1994).

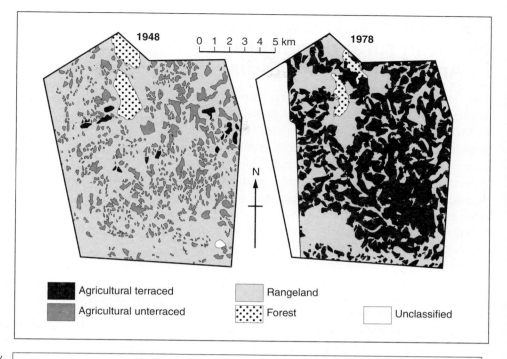

Legend:
- ■ Agricultural terraced
- ▨ Agricultural unterraced
- ░ Rangeland
- ⦂ Forest
- □ Unclassified

▶ **Figure 3.25** Community involvement. From C Lowe Morna, *The Power to Change*, Zed Books (1992) p. 47.

Africa abounds with examples of contour ridges, bunds and windbreaks, built under duress and, with no effort to involve the local community, that have simply been allowed to disintegrate. In Ethiopia, for instance, under the former socialist government of Mengistu Haile Mariam, peasant associations were obliged, in order to stop soil erosion, to construct thousands of kilometres of dry stone and earth walls called bunds. Millions of trees were also planted. Those who did not co-operate were fined. Today, there is precious little evidence of either the bunds or trees.

In Lesotho, notes a senior official, 'Huge amounts of money have been spent on mechanical measures, like making terraces and waterways. But we have realized there is one element missing. We have been ignoring the very people who are supposed to benefit from these measures.' Now, he maintains 'There is a real shift in thinking. We are concentrating on education, creating awareness and popular participation.'

1 a) Produce labelled drawings of the two main types of terrace construction to show water movement over and through the slope.
  b) What are the main advantages and disadvantages of each terrace type?

2 Study the photographs of the Machakos hillside (Figure 3.23). Use what you have learned about agroforestry and terracing to show how the physical and economic environments have been improved.

3 Use tracing paper and a ruler to draw five random lines across each map in Figure 3.24.

Use these to calculate the percentage of each land use in 1948 and 1978.

4 Study Figures 3.22 and 3.25. Evaluate the success of the three main periods of terrace building:
  a) colonial (1937–62)
  b) self-help (1965–present)
  c) government and NGO aided (1978–present)

5 How far have these measures made Malthus' predictions more or less likely?

Case study 3: The role of NGOs in rural development in Andhra Pradesh, India

The two Kenyan case studies have been academically researched. The data provided allow us to quantify and evaluate the effects of population pressure on the rural environment. They also indicate that colonial ideas were powerful in influencing rural management decisions and that some of these decisions may have excluded local people.

The third case study is rather different. It is located in another poor LEDC with a predominantly rural population. But the information is less quantified and more anecdotal. It illustrates the way in which an NGO has evolved while trying to help rural people to help themselves. It provides a very down to earth example of how communities can participate in their own development.

Andhra Pradesh is the fifth largest state in India. It is located in the south-east of the country and has a long coastline along the Bay of Bengal (Figure 3.26). The economy is dominated by farming, with 74 per cent of the population relying on the agricultural sector, although this sector only produces 41 per cent of the state's wealth. The state's farms produce 7 per cent of all of India's food grain and 40 per cent of the nation's spices, as well as significant quantities of oil seed, eggs and fruit. However, if we compare various socio-economic indicators, Andhra Pradesh appears to be somewhat below the national average. For example, infant mortality is 73 per 1000. This is marginally better than the national average of 80 per 1000, but far worse than other states, eg Kerala, where infant mortality is only 17 per 1000.

▼ **Figure 3.26** The location of Andhra Pradesh.

▼ **Figure 3.27** Literacy and wealth in Indian states.

State	Literacy rate (%)	State domestic product (rupees per person)
Andhra Pradesh	44	5570
Arunachal Pradesh	42	5551
Assam	53	4230
Bihar	39	2904
Goa	76	8096
Gujarat	61	6425
Haryana	56	8690
Himachal Pradesh	64	5355
Jammu and Kashmir	31	4051
Karnataka	56	5555
Kerala	90	4618
Madhya Pradesh	44	4077
Maharashtra	65	8180
Manipur	60	4180
Meghalaya	49	4458
Mizoram	82	no data
Nagaland	62	5810
Orissa	49	4068
Punjab	59	9643
Rajasthan	39	4361
Sikkim	57	5416
Tamil Nadu	55	5078
Tripura	60	3569
Uttar Pradesh	42	4012
West Bengal	58	5383
Average for India	52	5781

ARTIC: 'Instead of blaming darkness better light a candle'

ARTIC (Appropriate Reconstruction Training and Information Centre) is an NGO working in the northern region of Andhra Pradesh. ARTIC was founded in 1977 after a disastrous cyclone had caused widespread loss of life and damage along the northern coastline of the state. The original aim of ARTIC was to assist communities affected by natural disasters such as cyclones and flooding. The project aimed to help

1 a) Would you expect literacy to be linked to wealth? Justify your answer.
   b) Test the relationship by using Figure 3.29 to plot either a scattergraph or Spearman's Rank.

2 a) How well developed economically is Andhra Pradesh compared to other Indian states?
   b) What other data would help you to answer this?

rehabilitation of villagers affected by such tragedies, but also to help them make effective preparations for the next hazard.

During the 1980s and 1990s ARTIC has broadened the scope of its work. It now sees its main function as the support of the sustainable development of rural communities. The concept of sustainable development is explained in more detail in Chapter 5, page 68. ARTIC attempts to achieve this by:

- introducing technologies that are appropriate to the rural economy of this part of India
- involving rural communities in decision making and action (it currently works with around 70 different villages)
- acting as an intermediary between government officials and villagers by explaining government policies and becoming involved in health and social education as well as development training.

ARTIC states that its mission is to 'mobilize rural people to transform their villages into democratic self governing entities that promote equity, uphold social justice and inculcate self-reliance'.

ARTIC is managed by a governing body whose members are elected and responsible to the general body of ARTIC. The director also serves as the Secretary of the Society. He is assisted by a team of Programme Co-ordinators, Community Organizers and other staff. The staff of ARTIC are recruited from the project area. There are currently 75 persons serving ARTIC.

The staff of ARTIC are supported by the elected leaders of the Village Associations and functionaries drawn from the village.

ARTIC receives grants from OXFAM and ACTIONAID, two British NGOs. The society for the promotion of Wastelands Development, an Indian NGO, also supports ARTIC financially. In addition, ARTIC seeks the assistance of various government development agencies, such as the District Rural Development Agency and the Integrated Tribal Development Agency.

ARTIC achievements

Among its achievements ARTIC has:

- instituted Village Associations in 70 villages which are assuming responsibility for community development
- enrolled 3000 women in thrift and credit groups reducing their dependency on money lenders
- developed 261 hectares of fallow land
- planted 515 hectares of agroforestry
- enabled 516 families to acquire title to land by settling disputes, securing government assignment or aiding purchase
- constructed seven check dams, 34 tanks, 26 lift irrigation schemes, 215 dug wells and bore wells to provide irrigation which has led to more reliable crops and increased crop yields
- provided hand pumps for drinking water in 45 villages
- developed roads to 29 villages and provided electricity to 15 tribal villages
- increased enrolment of children for primary education and improved literacy levels of adult women and working children.

▲ **Figure 3.28** The provision of clean water is one of ARTIC's main aims.

- increased aid to mothers of small children to 92 per cent and child immunization to 88 per cent.

ARTIC's irrigation projects

Average rainfall in northern Andhra Pradesh is around 1100 millimetres per annum. It is intense and falls in an average of 70 days between June and October. The growing season therefore extends to no more than five or six months and agricultural labour is seasonal. Un- and underemployment is the norm during January to May. Furthermore, 50–60 per cent of all landowners have plots of land smaller than 0.2 hectare. ARTIC recognized that irrigation projects would increase rural wealth in two ways:

- construction of irrigation projects during the dry season would relieve the effect of seasonal employment
- water storage into the beginning of the dry season would increase the length of the growing season – if two crops can be grown a year then farm incomes could be doubled.

Access to clean water is a basic health issue. Most rural parts of Africa and South Asia have more limited access to clean water than urban areas. ARTIC have made the creation of a safe supply of water a major priority. Most of the 70 villages in which they work now have a secure water supply. The World Health Organization (WHO) recommend that there should be one safe water source for every 50 families. ARTIC's next target is to increase the number of wells in the larger villages.

ARTIC have worked with communities on a number of different irrigation projects which are listed below. Most irrigation projects are constructed during the dry season, thereby providing work for underemployed labourers.

- Bunds are low earth embankments which are built along the contours of fields. Bunds harvest rain water by reducing the rate of overland flow and encouraging infiltration, thereby recharging the soil moisture. Bunds also help to retain precious top soil.
- Tanks are larger earth embankments which are built as dams across streams to create temporary reservoirs of water. These tanks hold water during November to January or February, extending the season of available water by three to four months.
- Checkdams are small-scale concrete dams that serve a similar function to tanks.
- Pumpsets: diesel or electrically operated pumps are used to pump water out of mountain streams to irrigate fields.
- Borewells are drilled 30–60 metres deep. They provide clean drinking water in villages but ARTIC has found that they are less appropriate for farms. It has been found that overabstraction from a 60-metre borewell can lower the water table so that a neighbour's 10-metre deep well dries out. Only a few farmers have borewells, so control of the water resource is centralized in just a few hands. ARTIC is also critical of the construction and maintenance of the borewell. Much of the investment in a borewell is sucked out of the area. The drilling rig, the tube lining, the pump and generator are all manufactured outside the area. If the generator breaks down it requires expensive parts that are not made in the rural area as well as the expertise of an outsider to maintain the pump.

▶ **Figure 3.29** These photos show the manual construction of a shallow well lined with stone.

◀ Figure 3.30

ARTIC invested heavily in borewells in its early years, but now favours the construction of shallow wells.

- Shallow wells are built using manual labour. They are around 10 metres deep, have a diameter of 10 metres and are lined with local stone to prevent the sides collapsing. Water is raised using a pail. Local masons have been trained to make a foot operated pump which uses intermediate technology designed in the UK. The benefits of such wells are retained within the community because all the materials, labour, skills and training required are found locally.

1 How far do you think ARTIC's work is
   a) appropriate and
   b) sustainable?

2 Evaluate the role of ARTIC and other NGOs examined in this chapter. Assess their environmental, economic and social impacts. In what ways do they complement the role of government agencies?

3 Some social scientists consider that population growth is the major issue facing developing countries today. Others consider that it is also essential to consider the distribution of resources, including the distribution of political and economic power and decision making. Choose one of these viewpoints and justify it in the light of what you have read in this chapter.

Ideas for further study

What is the role of women in the rural development process in the LEDCs? You could start your investigation by visiting the Indian and African websites suggested. They should lead you to other websites which describe the role of women in development projects.

Summary

- Despite rural-to-urban migration, natural increase in the population has resulted in increasing rural population densities.
- Rural population growth in LEDCs does not necessarily lead to environmental crisis and deteriorating quality of life. There is evidence to suggest that environmental management practices such as agroforestry, and soil and water conservation techniques can actually enhance the rural environment and local incomes.
- Insensitive decisions by colonial and post-colonial governments have sometimes created rather than solved rural problems.
- The local community must be fully involved in decision-making if their efforts are to be successful in improving rural quality of life.

References and further reading

Andrew Goudie and Heather Viles (1997) *The earth transformed: An introduction to human impacts on the environment.* Blackwell.

Jane Chrispin and Francis Jegede (1996) *Population, resources and development.* Collins. (especially chapter 5)

Garrett Nagle and Kris Spencer (1997) *Sustainable development.* Hodder & Stoughton. (especially chapter 2)

A detailed description of the Machakos case study can be found in: ed. Tony Binns (199x) *People and Environment in Africa.* Wiley.

Ngũgî (1967) *A grain of wheat.* Heinemann.

The World Conservation Monitoring Centre has an excellent website providing detailed information on the conservation and sustainable use of the world's living resources. Their site can be visited at http://www.wcmc.org.uk

CPR Environmental Education Centre has a website on the Indian environment which can be viewed at http://www.cpreec.org

EcoNews Africa is an NGO initiative analysing global environment and development issues from an African perspective. Their website address is at http://www.web.apc.org/~econews

4 Rural deprivation and regional aid: A case study of the European Union

Introduction

In Chapter 1 you saw that image and reality can sometimes be quite different. Many perceive rural areas to be quiet places for rest and relaxation, whereas those who live there are more aware of the drawbacks. Lack of services, low wages and underemployment result in rural deprivation for some rural regions, especially those that are more isolated from major centres of population. This chapter returns to this theme, but on a European rather than a local scale. You will be investigating which parts of Europe can be defined as suffering rural deprivation. The second part of the chapter examines how the European Union (EU) is involved in managing its rural environments. Can the EU help create rural jobs and boost the rural economy? Can it narrow the gap between deprived rural regions and the wealthier and more vibrant economies of the European cities?

Rural regions of the European Union

The rural population of the European Union countries is around 86.5 million people, or 23.2 per cent of the total population of the fifteen countries of the EU. You saw in Chapter 1 that the rural population is a minority within UK society, and the same is true in other countries such as the Netherlands and Belgium. However, in other EU states the rural population is sizeable, as can be seen in Figure 4.1. The authors John and Francis Cole (1997) have estimated that between 96 and 97 per cent of the total land area of the EU can be considered rural. However, relatively few people are actively involved in managing this huge area of land. The Food and Agriculture Organization estimate that only 8.5 million people were actively involved in agriculture in the EU in 1994. This means that only 10 per cent of rural dwellers are farmers. The majority of the rural population are engaged in manufacturing or service industries, are retired or commute to work in urban areas.

▶ **Figure 4.1** Rural population of each member state of the EU.

	Total population, 1995 (millions)	Rural population, 2000 (%)
Austria	7.7	39
Belgium	10.1	2
Denmark	5.2	11
Finland	5.1	26
France	57.7	24
Germany	81.3	14
Greece	10.4	32
Ireland	3.6	36
Italy	57.1	28
Luxembourg	0.4	13
Netherlands	15.3	11
Portugal	9.9	61
Spain	39.1	17
Sweden	8.7	14
UK	58.2	6
EU total (15)	**369.7**	**23**

Different types of rural environment

Urban and rural areas are at different ends of a sliding scale known as the rural–urban continuum (see Chapter 1, page 8). Different rural areas also fall in different places on the continuum. Some are relatively close to large urban centres of population while others are extremely remote. Kent and Cambridgeshire in England, for example, contain rural landscapes and some parts have sparse populations. But these rural communities are close to major transport routes, work, shops, education and welfare facilities, and entertainment available in either London or one of its satellite towns. South Shropshire, studied in Chapters 1 and 2, is more remote and contains some communities which are deprived of services and isolated from larger towns and their facilities. But the most isolated rural communities in the UK are probably in the Highlands and Islands of Scotland. Rural communities in Europe face different issues depending on their proximity or isolation from the urban cores.

Rural environments close to urban cores

Figure 4.2 shows the location of cities in Europe with a population greater than 700 000. Some of these cities have larger populations, more commercial interests with more company head offices and, therefore, better paid workers than others. Many of the larger, most competitive and influential cities lie in an arc between London and Milan in Italy. Geographers have called this region the 'blue banana'. It is a region of power and prosperity although there is evidence that some cities in the region which are based on traditional industries may soon lose their dominance and the banana may begin to move southwards (see the

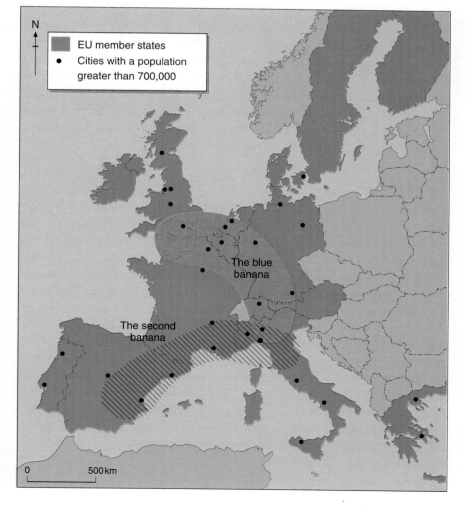

◀ **Figure 4.2** European cities with a population greater than 700 000 and the 'blue banana' (source: Drake, 1994, pp. 121–2).

'second banana' along the Mediterranean coast in Figure 4.2). Nevertheless, this banana-shaped region is an important influence on the European economy and is an example of a core region (see page 57).

Kent and Surrey in England, and the Pas de Calais and Picardie in northern France are rural regions that lie within the blue banana. These rural environments are within the sphere of influence of London and Paris. Counterurbanization is changing the demographic characteristics of rural environments that, like these, are accessible to large European towns and cities. Commuter traffic threatens the tranquillity of rural landscapes and changes the character of historic villages. Factories, office and retail units consume large areas of once green space. Patterns of employment are changed and incomes increase within the rural area. Second home owners and commuters buy up the limited number of country houses and cottages, inflating the cost of housing and changing the socio-economic characteristics of rural settlements. In the UK the sale of council houses in rural areas since The Right to Buy Act of 1981 has further reduced the stock of low-cost housing.

The effects of counterurbanization are not limited by national borders. In recent years the number of British buying second homes in rural areas of France has increased dramatically. The Dordogne, in the south-west, now has so many British expatriates that there are cricket clubs and pubs selling real ale. The price of second homes in the Dordogne rose by 10 per cent in 1997, while house prices generally in France remained static. The opening of the Channel Tunnel and the cut-price ferry war that followed has helped to make Normandy in the north of France another popular second home destination. Estate agents, like the one in Figure 4.3, appeal specifically to English-speaking clients.

▼ **Figure 4.3** An estate agent in Picardie, one hour from Calais, appeals to second home owners.

Rural areas on the edge of Europe

In the most remote rural areas of the EU the standard of living is often well below that enjoyed by urban dwellers. Many people in these remote districts have no mains water, sewage or gas. Many rural areas still have poor roads and inadequate public transport. Most significantly, perhaps, average wages are lower and job opportunities rare. It is not surprising, therefore, that in the most deprived rural areas, net out-migration has had a dramatic impact on demographic structure.

Rural population decline has had massive impacts on villages in the more mountainous regions of France and Spain, such as the Massif Central and Pyrenees. One such region is the Cevennes, in southern France. This sparsely populated and inaccessible region was once reliant on agriculture. Farmers kept sheep and goats. They cultivated the steep slopes with chestnut trees for their nuts and woodfuel, and mulberry trees on which they farmed silk worms. But at the beginning of the twentieth century two disasters ruined the fragile economy. The sheep and goats had been allowed to overgraze the mountainsides and pasture became scarce. At the same time disease killed the silk worms and the silk industry collapsed. Many people deserted their isolated villages, leaving some homes derelict to this day, as can be seen in Figure 4.4. The area is now a National Park, and tourism is the major industry. Even in this remote environment, second home owners are beginning to renovate the derelict properties.

▶ **Figure 4.4** Derelict properties in the villages of the Cevennes, southern France, are evidence of rural depopulation.

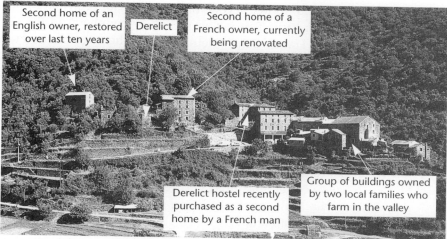

Second home of an English owner, restored over last ten years

Derelict

Second home of a French owner, currently being renovated

Derelict hostel recently purchased as a second home by a French man

Group of buildings owned by two local families who farm in the valley

Accessibility in the European Union

You have seen that rural areas may be divided between those that are accessible to major centres of population, and those that are more isolated. Accessible rural areas may suffer from the gradual sprawl of city life: they become less rural in character as more commuters and day trippers move in to enjoy the countryside. Remote rural areas suffer a different set of problems: lack of work leads to depopulation and a decline in essential services. Accessibility is, therefore, an important factor in determining the fortunes of different rural areas within the EU.

Figure 4.5 evaluates accessibility within the EU. It uses average travel times for heavy goods vehicles travelling from 18 major cities of the EU to each of the ten 10 per cent centroids of population within the EU. Each 10 per cent centroid is a point at the centre of population that comprises 10 per cent of the entire EU population. The centroids are located at Manchester, London, Paris, Cologne, Copenhagen, Munich, Bari, Milan, Toulouse and Madrid. Of these 10 centroids of population, Paris is the most accessible to the rest of Europe, with a mean travel time of just 13.9 hours to any other part of the EU. This travel time is expressed as an index of 100. Isolines have been plotted on the map as they relate to this index. For example, the 200 isoline passes through Madrid, indicating that average travel times from Madrid to anywhere else in the EU are twice as long as they are from Paris (i.e. a mean of around 28 hours).

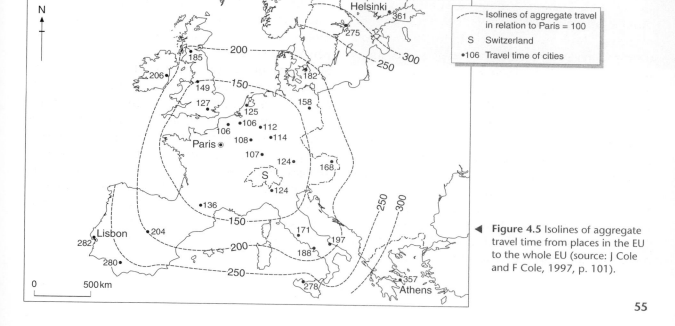

◀ **Figure 4.5** Isolines of aggregate travel time from places in the EU to the whole EU (source: J Cole and F Cole, 1997, p. 101).

The map on page 55 shows a core region of accessibility centred on Paris. This region includes south-east England, France, the Netherlands, Belgium, Germany, Austria and northern Italy. Rural areas within this region are likely to be most prone to the processes of counterurbanization. Rural regions on the fringe of Europe are much less accessible. Regions such as southern Spain, Brittany (France), Wales and Scotland, are all marginal or peripheral in the sense that they are physically at the edge of the European Union. But in some of these places at the edge of Europe the economy is stagnating too because of their isolation from the urban cores. So their economy is also marginal or peripheral to that of the rest of Europe.

1 Use Figure 4.5 to identify those parts of the EU which have a travel time index greater than 200. In what ways are these regions marginal to the rest of Europe?

2 What are the likely impacts of changes in travel time in marginal regions?

▼ **Figure 4.6** Away from the urbanized and densely populated coastal resorts of southern Spain the rural populations are extremely sparse. The rural economy is dependent on extensive practices such as the cultivation of olive trees as shown in this photograph. The EU has helped to fund major road schemes here to improve accessibility. Tourism and commerce should benefit as a result.

Improving access within the EU rural environment

Improvements in transport technology, new roads and wider car ownership have all combined to reduce travel times in Europe. The construction of the Channel Tunnel has further helped to reduce travel times from the UK to the rest of the EU. This has the effect of making rural places more accessible to the processes of counterurbanization. For example, more British second home owners can buy properties in France. Improved accessibility has the effect of making the world seem a smaller place. Distances and differences between places seem to diminish as transport and communications improve. The 'shrinking' of the globe in this way is known as time–space convergence.

Politicians within the EU have recognized that improving accessibility could be the key to unlocking the stagnating economies of parts of peripheral Europe. Many European rural environments are very beautiful and, if made more accessible, could benefit from the growth of the tourist economy. The economic aid given by the EU is discussed in more detail on page 61, but it is worth noting here that millions of ECU have been spent on improving roads in deprived rural regions such as Andalucia, southern Spain. The photograph in Figure 4.6 shows the beautiful scenery (only a few kilometres from the Costa del Sol) which is now accessible to the tourists.

Core and periphery in the European Union

You have seen that some regions of the EU are more accessible to the rest of Europe than others. Evidence suggests that the more accessible regions have economic advantages over other parts of the EU, making them economic cores. Other regions are physically and economically more isolated. These are the marginal or peripheral regions. The concept of core and periphery regions is discussed more fully on pages 58–59.

Evaluation of core and peripheral regions requires the use of quantified economic criteria.

Data on subjects such as unemployment, or the percentage of the workforce employed in agriculture, manufacturing or service industries can be used as economic evidence of a region's importance. Figures 4.7 and 4.8 use three criteria to compare the relative strength and weakness of regions within the EU. Figure 4.7 maps the average contribution of each region to national wealth (GDP). Figure 4.8 provides data on unemployment and the percentage of the workforce employed in agriculture.

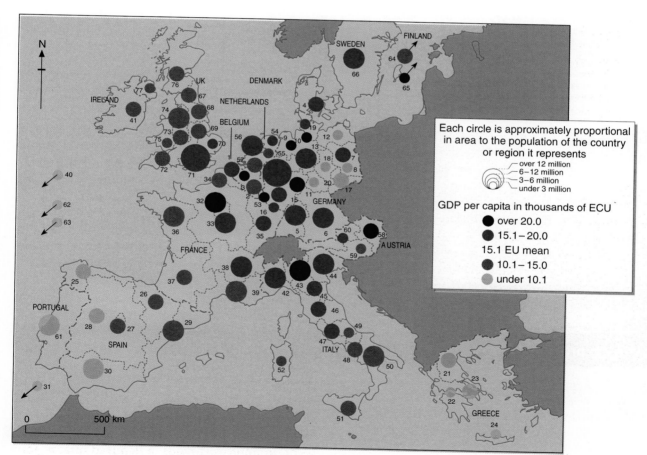

▲ **Figure 4.7** Regional variations in GDP per capita within the EU, 1995 (source: J Cole and F Cole, 1997, p. 272).

▼ Figure 4.8 Regional variations in unemployment within the EU, 1995 (source: J Cole and F Cole, 1997, p. 380–2).

Region	Employed in agriculture %	Unemployment %
Belgium	2.6	8.2
1 Vlaams Gewest	2.7	6.1
2 Région Wallonie	3.1	11.2
3 Bruxelles–Brussel	0.2	10.8
4 Denmark	5.1	10.3
Germany	3.6	7.5
5 Baden–Württemberg	3.2	4.2
6 Bayern	5.8	4.0
7 Berlin	0.7	10.4
8 Brandenberg	4.7	14.0
9 Bremen	0.8	9.1
10 Hamburg	1.2	6.1
11 Hessen	2.6	4.9
12 Meckelnburg–Vorpommern	7.8	16.9
13 Niebersachsen	4.8	6.7
14 Nordrhein Westfalen	2.1	6.9
15 Rheinland–Pfalz	3.5	4.9
16 Saarland	0.7	7.7
17 Sachsen	3.1	13.7
18 Sachsen–Anhalt	4.5	15.9
19 Schleswig–Holstein	4.7	5.5
20 Thüringen	3.3	15.0
Greece	21.3	7.5
21 Voreia Ellada	30.5	7.3
22 Kentriki Ellada	38.5	7.9
23 Attiki	1.1	10.4
24 Nisia	28.2	4.3
Spain 21.5		10.2
25 Noroeste	24.7	18.7
26 Noreste	7.3	18.8
27 Madrid	0.9	16.9
28 Centro	16.5	21.3
29 Este	5.2	19.2
30 Sur	13.3	30.1
31 Canaries	7.5	26.9
France	5.9	11.1
32 Ile de France	0.5	9.6
33 Bassin parisien	8.9	11.6
34 Nord-Pas-de-Calais	3.2	14.5
35 Est	3.7	8.8
36 Quest	10.4	10.8
37 Sud-Quest	11.0	11.2
38 Centre-Est	6.0	10.8
39 Mediterranée	4.4	14.6
40 Départements d'Outre-Mer	7.2	n.a.
Ireland	12.8	15.4
Italy	7.3	10.4
42 Nord Ouest	5.5	7.9
43 Lombardia	2.8	5.3
44 Nord Est	6.4	5.4
45 Emilia–Romagna	6.9	5.7
46 Centro	6.1	7.5
47 Lazio	3.5	10.2
48 Campania	10.0	20.2
49 Abruzzi–Molise	12.5	10.9
50 Sud	15.4	16.0
51 Sicilia	13.4	19.2
52 Sardegna	13.2	18.9
53 Luxembourg	3.2	10.4
Netherlands	4.0	6.5
54 Noord-Nederland	5.3	8.3
55 Oost-Nederland	5.4	6.2
56 West-Nederland	3.2	6.3
57 Zuid-Nederland	4.1	6.1
Austria	7.0	3.9
58 Ostösterreich	n.a.	4.3
59 Südösterreich	n.a.	4.1
60 Westösterreich	n.a.	3.4
Portugal	11.6	5.2
61 Continente	11.4	5.3
62 Açores	18.0	5.0
63 Madeira	14.3	3.9
Finland		
64 Manner-Suomi	9.0	16.4
65 Ahvenanmaa	n.a.	5.1
66 Sweden	3.0	7.9
United Kingdom	2.0	10.0
67 North	1.8	11.3
68 Yorkshire and Humberside	1.7	9.9
69 East Midlands	2.4	8.8
70 East Anglia	4.5	7.7
71 Southeast	1.2	9.9
72 South West	3.6	9.0
73 West Midlands	1.6	10.4
74 North West	1.0	10.4
75 Wales	3.7	9.6
76 Scotland	3.1	9.9
77 Northern Ireland	4.8	15.0
EU total	5.6	10.4

1 In groups of two or three show whether it is possible to identify a 'core' of the EU using Figure 4.7. Justify your answer.

2 Compare the pattern in Figure 4.7 with the data in Figure 4.8.
   a) To what extent is it possible to identify regions that show evidence of deprivation using all three criteria?
   b) What is the correlation between the more isolated parts of Europe shown in Figure 4.5 and economic deprivation?

3 Assess the use of the three criteria. What other criteria could be used to identify core or peripheral regions within the EU?

Core and periphery theory: convergence or divergence?

Core and periphery theories are those which deal with the differences between richer and poorer regions. Early theories simply attempted to describe the differences between regions, while later more complex theories attempt to explain the dynamics of change.

Core economic regions are those areas which are characterized by:

- high levels of employment and a largely skilled workforce
- investment and sustained economic growth.

Core regions attract investment in new growth industries. New jobs are created so local people have more money to spend and invest. This in turn creates more wealth for other local businesses such as shops and banks. The economy of the core region spirals upwards because of these multiplier effects.

Peripheral economic regions are those areas characterized by:

- high levels of unemployment and a largely unskilled or deskilled workforce
- lack of investment and economic decline.

Peripheral regions have few employment opportunities. Wages are lower so personal wealth and investment are also lower. This means there is little money circulating through the shops, banks and other service industries of the local economy. There is a danger that the economy will go into a spiral of decline unless new investment is put into creating jobs or training people.

Core–periphery flows

Core and periphery regions do not exist in isolation from one another. People, goods and money (capital) will flow from one region to another, as shown in Friedman's dynamic core–periphery model in Figure 4.9.

Businesses based in core regions may invest money in new factories or offices in peripheral regions. New jobs may begin to create multiplier effects in the local economy of the peripheral region. The investment may create training opportunities and therefore improve employment opportunities. Flows such as this which benefit the peripheral region are termed spread effects. The promise of spread effects encourages the funding of regional aid projects in the poorer regions of the EU. European ministers expect that the benefits of investment will spread or 'trickle down' to aid the wider economy. Spread effects should help the economy of the peripheral region to catch up with the core region, a concept known as convergence.

However, flows of people, goods and services also move from the peripheral to the core region. The most mobile people leave the declining region to find better jobs in the core. Raw materials, such as foodstuffs or minerals, are exported from the periphery often in an unprocessed form. They are then processed in factories in the core, which adds considerable value, before being sold to consumers in both regions. For example, LEDCs in the Pacific export foodstuffs such as pineapples or coconut to TNCs, but import basic foodstuffs from MEDCs, causing balance of trade problems. Flows from periphery to core are known as backwash effects or leakage. The result of leakage is that the rich region gets richer, while the periphery gets relatively poorer. Divergence is the term used to describe the widening of the economic gap between rich and poor.

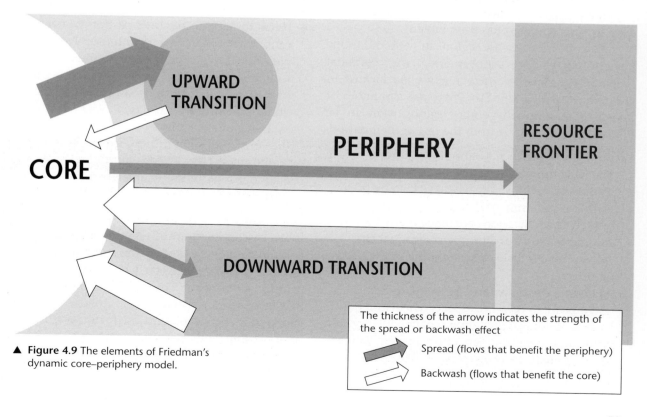

▲ **Figure 4.9** The elements of Friedman's dynamic core–periphery model.

The thickness of the arrow indicates the strength of the spread or backwash effect

Spread (flows that benefit the periphery)

Backwash (flows that benefit the core)

National separatism within the EU

The last thirty to forty years has seen the development of apparently conflicting ideals within the European Union. The EU currently consists of 15 member states. It is an example of a decision-making body that creates transnational policies on many issues, including the rural economy. The main aim of the EU is to create a large, stable region in which inequalities are removed. But while the EU has been developing closer co-operation and integration between its member states, a number of regions within the EU have begun to campaign for greater independence, a movement described as national separatism. Examples can be seen within the UK, where Scottish nationalists have campaigned successfully for their own tier of national government. In other parts of Europe, regions which consider themselves to have a distinct culture and heritage of their own have struggled to become separate from their own state. Brittany, for example, in western France, has a culture and language that has close connections to the other Celtic regions of Europe: Ireland, Wales and Scotland. Figure 4.10 shows Breton people carrying their flag at a folk event. In other regions, for example the Basque region which includes part of France and Spain, the national separatism campaign has been violent.

Figure 4.11 locates the regions in western Europe where there are movements for national separatism. Many of these regions are located in marginal parts of the EU. There are a number of possible reasons why some regions want to be recognized as separate from the rest of their nation:

- Some movements for national separatism reflect the need for greater economic development. Peripheral regions may feel neglected by their national government. Local people may perceive that decisions are made many kilometres away in a prosperous capital city where there is little understanding of the needs of others.
- Many national separatist movements are attempting to preserve aspects of a distinctive cultural heritage, including separate history, art, music and language. Separatists may be seen as trying to halt the progress of globalization by keeping their own region distant and unique.

▼ **Figure 4.10** Folk musicians and dancers in Brittany celebrate a culture and language separate from the rest of France and proudly carry the Breton flag.

- It was thought in the early part of the twentieth century that there was a minimum size for any nation. Below a certain size a nation's economy would be too small to support itself, especially its welfare and defence needs. However, the development of the EU has allowed relatively small regions to consider independence within the context of support from the wider European community. Thus, Plaid Cymru, the Welsh Nationalist party, has campaigned for 'full self government for Wales within the EU'.

You saw in Chapters 1 and 2 that people in UK rural environments are attempting to preserve their cultural heritage. Those living in rural areas may feel marginalized, neglected or misunderstood by central government. The rural economy may also be stagnating. There are, therefore, many similarities between separatist movements and rural campaigns. Several regions currently campaigning for separatism are marginal rural environments.

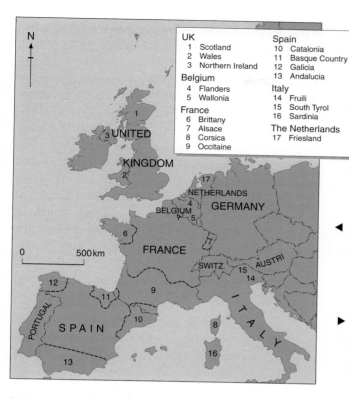

UK
1 Scotland
2 Wales
3 Northern Ireland

Belgium
4 Flanders
5 Wallonia

France
6 Brittany
7 Alsace
8 Corsica
9 Occitaine

Spain
10 Catalonia
11 Basque Country
12 Galicia
13 Andalucia

Italy
14 Fruili
15 South Tyrol
16 Sardinia

The Netherlands
17 Friesland

1 Compare Figure 4.11 with Figures 4.2, 4.5 and 4.7. Which regions claiming some form of separatism are also:
   a) in inaccessible parts of the EU
   b) relatively deprived
   c) rural as well as physically and economically marginal?

2 What conclusions can you draw from your answer to question 1?

◀ **Figure 4.11** Regions of Europe in which nationalists claim some form of independence (source: Anderson, Brook and Cochrane, 1995, p. 91).

▶ **Figure 4.12** Assistance within the EU via the CAP and regional aid (Structural Funds) in billions of ECU.

	CAP	Structural Funds
1987	32.7	9.1
1997	35.6	29.3

Narrowing the urban–rural gap in the European Union

Decisions about the management of Europe's rural regions are made by a huge number of different individuals and groups. Only some of these groups work in partnership with each other. Many other groups have differing points of view and consequently there are often conflicts of opinion. Many of these groups operate locally within the rural area, for example landowners, farmers and local conservationists. Other interested groups work at a national level, such as government organizations and NGOs, for example the National Farmers Union (NFU). In Europe there is a third and even wider level of decision making. The European Union currently consists of fifteen member states. The main aim of the EU is to create a stable macro-region in which inequalities are removed. In order to achieve this the EU has created a huge bureaucracy in which decision making for all fifteen states is centralized.

Assistance for the deprived rural areas of the EU

It has been long recognized that Europe's rural areas and its farmers need financial assistance if convergence is to occur between the rich and poor regions of the EU. Assistance has been given for over three decades in the form of various subsidies to support agriculture, and grants to create new training and business opportunities. This assistance can be divided broadly between aid given directly to farmers through the Common Agricultural Policy (CAP) and regional assistance given to those areas that are suffering some form of deprivation. The proportion of assistance given through these two mechanisms has shifted towards regional assistance in recent years (Figure 4.12).

Common Agricultural Policy (CAP)
The CAP was first established in the late 1960s. The policy had several aims:

- to increase farm productivity and ensure that Europe was more self-sufficient in food
- to provide a decent standard of living for farmers
- to stabilize market prices for agricultural products by reducing competition and thereby preventing inflated prices from affecting consumers.

Each member state of the EU contributes money to support the aims of the CAP. The details of which farmers receive money and how much they receive has had to be renegotiated many times. But, in essence, the CAP is achieved by creating a balance between:

- subsidies of various amounts that are paid to farmers for producing particular agricultural products
- quotas that are placed on specific agricultural products to limit production in any one region.

The amount of money spent by the EU on the CAP has risen year by year. Farm production has increased so that the EU is now self-sufficient in most foodstuffs (although it is also the world's largest importer of agricultural products). By the mid-1980s the UK government began to complain that spending had got out of control. They argued that the UK's contributions to the CAP were greater than the benefits received by British farmers. The public no longer saw the need for increasing food supply. Many were angry that taxpayers' money was spent creating mountains of food that no one actually needed, especially when people in other parts of the world were hungry (Figure 4.13).

Attitudes in Germany towards the CAP were also shifting in the 1980s. The political success of the Green Party highlighted the fact that CAP money was frequently spent on intensive farming activities. These included increasing field sizes to make mechanization easier, and the greater use of herbicides, pesticides and fertilizers, all of which have a negative impact on the environment. However, the EU continued to support the main idea underlying the CAP: that economic needs were greater than environmental concerns.

In 1990 the disagreement over CAP spending reached crisis point. Nations outside the EU, such as the USA, complained that the CAP subsidies and quotas gave European farmers an unfair advantage on the world market. They accused the EU of having protectionist policies – in other words, of protecting EU farmers against cheaper imports. International agreement on trade was finally reached in 1993, and the need for reform of the CAP had been accepted. Other factors had helped to focus the minds of the European politicians:

- the reunification of Germany in 1990 – the wealthy former West Germany could no longer afford massive payments to the CAP because it had to support the much weaker East German economy
- the hardening of public opinion against modern farming techniques – various food scares, including the 'egg scare' (fear of salmonella) and 'mad cow disease' (BSE and its link to the human form, CJD) have made the public wary about the intensive way in which animals are fed and reared
- evidence that some farmers fraudulently claim subsidies, for example in Corsica in 1994, when the number of pigs farmed in the region was grossly exaggerated.

The CAP was reformed in 1992. Subsidies were reduced and a range of agri-environmental measures were introduced. But there was also a recognition that by the year 2000 the CAP would have to be radically changed. Many farmers in the UK are now most concerned about the uncertain future of their industry. The reduction in subsidies and the BSE crisis have ruined the incomes of many farmers in the poorer farming regions of the UK, such as south-west Shropshire (see Chapter 1).

EU Structural Funds and regional assistance

It was estimated in 1995 that about 85 per cent of the EU at that time was eligible for some assistance (Figure 4.15). This area contained only about 70

▲ **Figure 4.13** Negative perceptions of the CAP (source: *Thin Black Lines,* Development Education Centre, Birmingham (1988)).

▲ **Figure 4.14** Roadside banners encouraging the public to support British farmers sprang up in many rural areas in early 1998.

1 Analyze the figures in Figure 4.12 and comment on the impact the change in emphasis might have on the economy and environment of different rural areas.

2 In this section work in pairs to consider the CAP and all forms of assistance.
   a) Identify what you consider to be the benefits of assisting rural areas.
   b) Identify the problems.

3 Should rural areas be seen as 'special cases' for assistance?

per cent of the population of the EU because the more densely populated regions of the core do not qualify for assistance. The EU budget is equivalent to only 1 per cent of the total GDP of the EU. Of this budget, about a quarter is directed to assisting deprived regions.

The European Structural Funds are allocated to fulfil six priority objectives. Objectives 1, 2, 5b and 6 direct aid to particular regions. The other objectives direct funds to sectors of the economy or workforce although they indirectly benefit the same regions. The objectives are:

- Objective 1: To promote development in regions that are lagging behind.
- Objective 2: To redevelop regions seriously affected by industrial decline.
- Objectives 3 and 4: To combat long-term unemployment and to assist workers to adapt to industrial change.
- Objective 5a: To modernize production, processing and marketing structures in the agricultural, forestry and fishing sectors.
- Objective 5b: To promote development in rural areas which suffer sparse populations and declining services.
- Objective 6: To assist in the development of rural areas in Sweden and Finland north of latitude 60°N.

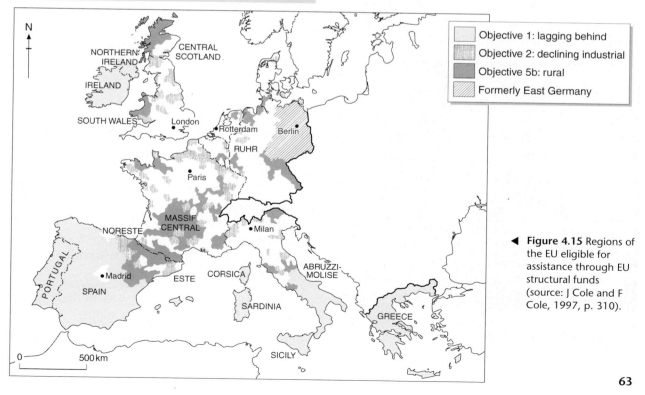

◄ **Figure 4.15** Regions of the EU eligible for assistance through EU structural funds (source: J Cole and F Cole, 1997, p. 310).

	Allocation (%)
Objective 1	65.7
Objective 2	4.9
Objectives 3 and 4	9.7
Objective 5a	3.8
Objective 5b	4.3
Objective 6	*

◄ **Figure 4.16** Allocation of Structural Funds between the six objectives (1994–9).

*Objective 6 was added after the addition of Sweden and Finland to the EU in 1995. Spending did not total 100 per cent because some funds remained unallocated at the beginning of the five-year period (1994–9).

	Total	Per capita	Objective 5b
Belgium	1859	182	77
Denmark	767	148	54
Germany	20 586	252	1 227
Greece	15 066	1 435	–
Spain	32 810	839	664
France	12 750	219	2 238
Ireland	6 004	1 668	–
Italy	20 679	358	901
Luxembourg	89	223	6
Netherlands	2 084	134	150
Austria	1 623	200	n.d.
Portugal	15 396	1 555	–
Finland	1 704	334	n.d.
Sweden	1 420	160	n.d.
UK	10 265	175	817
EU total (15)	143 102	384	6 134
EU percentage	100¹		4.3

◄ **Figure 4.17** Total assistance and area 5b assistance for each member of the EU (source: J Cole and F Cole, 1997, p. 311).

▼ **Figure 4.18** From J Cole, 1997.

Economic indicators for the regions of the EU show that in some member States the disparity in income has increased in the early 1990s. Until the 1980s, southern Italy and (after 1973) Ireland were the only two areas in the EC that lagged far below the average level. The addition of Greece, Spain and Portugal in the 1980s and of East Germany in 1990 has greatly increased the number of relatively poor areas to be assisted. In the 1950s and 1960s, economic growth in the EEC was rapid and it was more easy to transfer funds to poorer areas. With economic uncertainty in the 1970s and 1980s and a major recession in the early 1990s the equalization process is evidently faltering.

Objective 5b regions

▶ **Figure 4.19** From F Harrington, 'The ESF and the promotion of rural development under Objective 5b', Social Europe, 2/91.

Objective 5b was designed specifically to encourage a positive approach in the regions worst hit by agricultural decline and the reform of the Common Agricultural Policy by helping them to diversify, develop and revitalise their economies.

Objective 5b regions were first designated in May 1989. Eligibility depends on one or more of certain criteria. Eligible regions:

- have a large percentage of the population working in agriculture
- suffer depopulation due to out-migration
- have many farmers affected by the withdrawal of CAP subsidies
- are isolated from major centres of activity
- suffer specific problems because of their mountainous terrain.

In 1991 17 per cent of the land area of the EU became eligible for Objective 5b status although this area contained only 5 per cent of EU population.

The European Structural Fund is divided between three separate funds:

- the European Regional Development Fund (ERDF)
- the European Social Fund (ESF)
- the European Agricultural Guidance and Guarantee Fund (EAGGF).

Funding is available from all three of these funds to areas which have Objective 5b status.

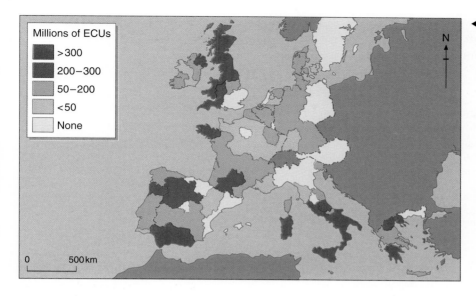

LEADER II projects

LEADER II is a Community Initiative funded by all three parts of the Structural Fund. It is committed to allocate around 1755 million ECU over its six-year programme (1994–9). Objective 1 (development lag), Objective 5b (rural) and Objective 6 (Nordic) regions are all eligible. The objectives of LEADER II are to:

- encourage local initiatives that may act as model developments to other rural communities in other parts of the EU
- increase exchanges of experience and transfers of know-how across the EU
- support transnational development projects.

LEADER II assists 'Local Action Groups' (combined public and private partnerships), local authorities, cooperatives, or any other non-profit making organization working in rural areas. The initiative supports a wide range of rural developments including:

- vocational training
- development of rural tourism
- increasing the commercial value of agricultural and fisheries products
- improved marketing of rural products.

LEADER II has created a 'European Rural Development Network' so that information about innovative projects can be shared through a series of meetings and publications.

LEADER II: A case study of Andalucia, Spain

▲ **Figure 4.21** The reservoir of La Vinuela (in the centre of Figure 4.22) supplies water to Malaga and the tourist resorts of the Costa del Sol.

La Axarquia, close to Malaga in Andalucia, is a LEADER II project area. This mountainous region was depopulated during the 1960s when people left to work in the new tourist developments along the Costa del Sol (Figure 4.21). The population has not recovered. Average age is increasing and the birth rate is falling. Agriculture is the major employer. There are two types:

- horticulture – fruit and flowers are grown on irrigated land, but there are problems with profitability (obsolete farming methods) and marketing

- traditional agro-food products (goat's cheese, oil, wine, raisins) are produced on the high slopes.

▼ **Figure 4.22** Factfile: La Axarquia, Andalucia.

La Axarquia, Andalucia

Surface area	1300 km²	Unemployment	30% (Spain=13.4%)
Population	47 000	Age of population	
Population density	36 inhabitants/km²	<25	30%
Working population		>64	17% (Spain=11.8%)
	agriculture 60%		
	industry 5%		
	services 35%		

▼ **Figure 4.23** The Malaga region.

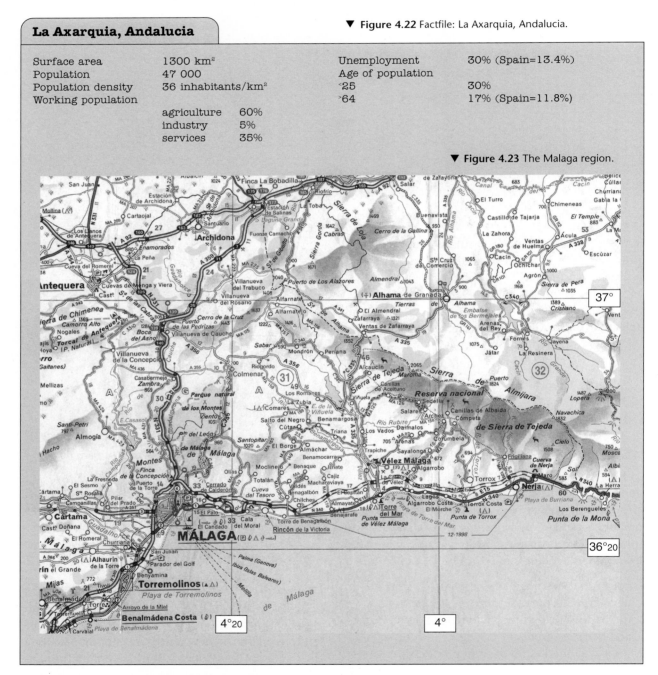

The LEADER II aid is being used primarily to develop rural tourism and to modernize the handicraft industry (ceramics, leather, wickerwork, esparto, equine products, furniture). LEADER II has committed 6 200 000 ECU, of which 2 611 000 comes from the EU, 2 326 000 from private sources and 1 263 000 from the state. The largest financial commitment is to developing rural tourism (3 949 000 ECU), followed by supporting agricultural production (1 008 000 ECU).

1 Study Figure 4.22. The region which receives LEADER II assistance lies in a triangle between Malaga and Nerja on the coast, and the Sierra del Alaljara mountains.

a) Justify La Axarquia's designation for LEADER II assistance.

b) Produce a sketch map of the region. Label the links between rural and urban areas in this region (consider people, investment, water and agricultural products).

c) Design a strategy that will attract tourists staying in Torremolinos to visit the rural tourist projects in La Axarquia.

Ideas for further study

- Investigate the proposed enlargement of the European Union. Which countries would like to join the current 15 members of the EU? What are the rural environments of these countries like? What issues would be faced by an enlarged EU?
- Do you consider that rural environments in the UK benefit from EU membership?

Summary

- Rural areas fall on a continuum between those that are accessible to large urban centres and those that are very isolated.
- Within the EU the most accessible rural environments are found within the core or 'blue banana'. Rural communities in this region are changing fast as a result of the counterurbanization process.
- The more marginal and less accessible rural environments of the EU are suffering various forms of deprivation. Some are suffering population loss. In others the economy is stagnating and needs assistance.
- There is a correlation between rural deprivation and national separatist movements in some marginal regions of the EU.
- EU assistance to rural communities is currently undergoing a painful period of change. Farmers are concerned about their future livelihoods as CAP is withdrawn. In the worst case scenario, rural economies would be devastated and the landscape could become derelict as farmers go out of business.
- EU assistance to areas eligible for 5b status is largely for diversification within the rural economy. Sustainable tourism projects are commonly used.

References and further reading

A huge mass of statistics on the European Union can be accessed at the Eurostat website at http://www.europa.eu.int/en/comm/eurostat
Leader II projects have a website at http://www.ruraleurope.aeial.bc

Useful books include:

J Cole and F Cole, *A Geography of the European Union,* 2nd ed, Routledge, 1997. (Some of the maps in this chapter have been sourced from this book.)

G Drake, *Issues in the New Europe,* Hodder & Stoughton, 1994.

J Anderson, C Brook and A Cochrane (eds), *A Global World?* Oxford University Press, 1995.

J Cole, *Geography of the World's Major Regions,* Routledge, 1996.

Managing rural tourism and recreation: a return to South Shropshire

Introduction

South Shropshire was introduced in Chapter 1 to challenge our image of the rural environment. This final chapter returns to the area to see how national and European policies affect people there. It picks up the theme of recreation in the rural environment that was examined in Chapter 2. You will see how national agencies affect local communities, and in particular, how the Countryside Commission works in partnership with other agencies to manage the Shropshire Hills Area of Outstanding Natural Beauty (AONB). As in the case studies of Kenya and Andhra Pradesh in Chapter 3, it is the efforts of local people which make national decisions work on a local scale. The chapter also looks at how the EU, described in Chapter 4, can assist local rural initiatives. By the end of the chapter, you will be in a position to make management decisions about the development of sustainable tourism in the Shropshire Hills through a decision-making exercise. What kinds of development would you encourage?

The Shropshire Hills AONB

The Shropshire Hills is an upland area of great beauty on the border between the West Midlands and mid-Wales. It was designated an AONB in 1959 (Figure 5.7). It was therefore one of the first AONBs, and remains one of the largest, covering 804 square kilometres. The landscape of the AONB owes its quality to the distinctive hills and ridges, as well as the mosaic of field patterns, hedges and hedgerow trees which are a product of centuries of farming. Agriculture still accounts for a large proportion of jobs in the area. However, the beef scare over BSE and changes to the CAP (see page 61) are leading to significant losses in the rural economy. The loss of jobs and income, and the loss of rural services in the relatively sparse and scattered communities of the Shropshire Hills are two of the most important issues facing the area.

One strategy used by local farmers and others has been to diversify their business into tourism. Information on visitor trends in South Shropshire are given in Figures 5.1 to 5.3. Leisure events such as beer festivals, summer carnivals and street fairs are now a regular part of the annual calendar in many of the small settlements of the Shropshire Hills. Meanwhile, farmers convert barns or stables into holiday accommodation. But the growth of the tourist industry in rural areas has its critics. The Country Landowners' Association (CLA) is concerned about allowing greater access to walkers, believing that its members should be allowed to enjoy their land in privacy and without the costly maintenance of footpaths and stiles.

Conservationists are worried too, about the environmental impact of rural recreation. Walkers trample vegetation and cause footpath erosion, scarring the landscape (see Chapter 2, page 21). Most visitors arrive by road, causing traffic congestion and pollution. These pressures are already evident in National Parks, especially the Lake District and Peak District. Could the same problems occur in the Shropshire Hills? The management problem is to find a way of encouraging economic development without damaging the fragile rural environment. One strategy is to encourage sustainable development projects (see page 71).

▼ **Figure 5.1** The distribution of tourist attractions in South Shropshire.

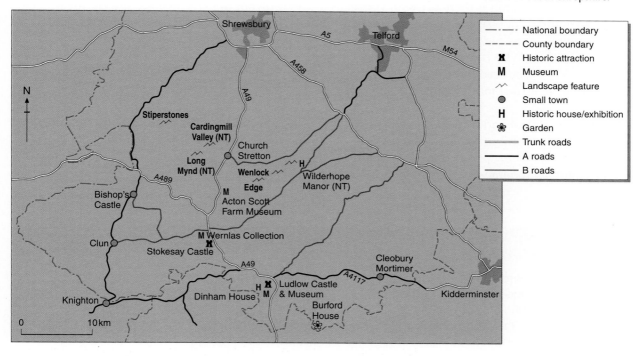

◀ **Figure 5.2** Responses to a survey of visitors to South Shropshire (source: 'A tourism strategy and action programme for South Shropshire', Shropshire Sustainable Rural Tourism Project).

Attraction	Type of attraction	% of survey who recognized the name	% of survey who had visited the attraction
Ludlow Castle	historic	92	21
Church Stretton	town	81	23
Wenlock Edge	landscape	70	8
Longmynd	landscape	65	14
Stokesay Castle	historic	63	13
Clun	town	58	5
Bishop's Castle	town	57	6
Cleobury Mortimer	town	55	3
Knighton	town	52	4
Ludlow Museum	historic	45	5
Cardingmill Valley	landscape	45	9
Acton Scott Museum	farming museum	44	5
Stiperstones	landscape	43	5
Dinham House	exhibitions	26	3
Burford House	garden	24	2
Wilderhope Manor	historic	21	1
Wernlas Collection	wild fowl	21	2

▶ **Figure 5.3** Breakdown of tourist expenditure in South Shropshire by visitor type, 1992 (%) (source: 'A tourism strategy and action programme for South Shropshire', Shropshire Sustainable Rural Tourism Project).

	Day	Touring	Overnight	All
Accommodation	–	–	52	23
Eating and drinking	36	36	22	30
Food shopping	4	11	4	6
Other shopping	37	24	11	21
Fuel and transport	9	15	6	10
Entertainment	14	13	4	9
Other	0	1	1	1

1 Study Figure 5.2.
  a) What correlation is there between visitor recognition of place names and the actual number of people who visit each attraction? Why is there this correlation?
  b) Which places need to promote themselves better to attract more visitors?

2 Study Figures 5.1 and 5.2.
  a) Trace the distribution of attractions in Figure 5.1. Draw proportional symbols on your map to represent the number of visitors to each attraction.
  b) Describe the distribution of visitors shown on your map. Which parts of the area have the most and which have the least visitors? Suggest reasons for this.

3 a) Study Figure 5.3. Summarize the main patterns of expenditure shown in this table.
  b) How should local businesses try to attract tourists to spend more locally?

What is sustainable development?

Sustainable development has been defined as 'development that meets the needs of the present without compromising the ability of future generations to meet their own needs' (source: 'Our Common Future', The Bruntland Report – Report of the 1987 World Commission on Environment and Development).

Many economic developments in the past have been concerned with maximizing profit with no real consideration for the impact on either the welfare of local people or the environment. By contrast, sustainable growth should allow economic benefits to percolate through the local economy without having lasting effects on the environment. Sustainable development should allow for improved standards of living and the maintenance of a healthy environment for future generations.

At the Rio de Janeiro Earth Summit of 1992, world leaders signed an agreement to implement local plans for sustainable development across the globe.

These local plans require the commitment and co-operation of people at grass roots level. The intention is to establish partnerships between the private, public and voluntary sectors to formulate local-scale, sustainable development strategies. Such strategies require:

- a period of consultation and auditing
- a plan of action which involves different sectors of the community
- a method of evaluation.

Many parts of the private sector have been quick to see the potential benefits of co-operation on local sustainable projects. Real savings can be made in such areas as water and energy use, and by the reduction of waste products. It is essential to acknowledge that sustainable development is about achieving economic, welfare and environmental benefits, not just for the moment but in the long term.

Who manages the Shropshire Hills?

The Countryside Commission (see page 21) has responsibility for designating AONBs. Unlike National Parks, AONBs do not have special Boards or special funding. Management decisions within the AONBs are taken by landowners, the local authority, and any government agencies (such as English Nature) or NGOs (such as the National Trust), which are working in the area. The Countryside Commission recommends that all government and voluntary agencies operating within an AONB should work together in partnership with one another. Such a partnership is known as a Joint Advisory Committee (JAC). According to the Countryside Commission, 'It is for the JAC to provide the impetus, the motive force and the leadership in co-ordinating the effective management of a nationally designated area'.

Government agencies working in the Shropshire Hills

English Nature is a major national agency working on conservation strategies in the Shropshire Hills. It has two roles:

- Management of National Nature Reserves established by the 1968 Countryside Act. The Shropshire Hills contains one Reserve, the Stiperstones, and parts of two others. The Stiperstones is an upland moor which is relatively small, just 400 hectares, and is easily accessible by road. The moor attracts many walkers and may be regarded as a honey pot site (see page 26). The condition of its footpaths has deteriorated in recent years (Figure 5.4).
- Selection and notification of Sites of Special Scientific Interest (SSSIs). There are numerous sites within the AONB, some with distinctive flora and fauna, others with geological interest. SSSIs are not protected by law, nor always accessible to the public.

◀ **Figure 5.4** Trampled vegetation and eroded footpaths on the Stiperstones moor. Bullet points below the photo explain the factors that contribute to moorland erosion.

- The harsh physical environment means that moorland vegetation is slow growing. Damaged plants do not have time to recover before being trampled again.

- The acidic, peaty soil is waterlogged in the winter, but can become very dry in the summer.

- The bedrock is an impermeable, resistant sandstone called quartzite.

- When the plants have died back, the thin soil is easily eroded by heavy rainfall and by walkers' boots.

- Walkers avoid the rutted, eroded path. The remaining soil is sticky and heavy to walk through, and the jagged rock beneath is very uncomfortable underfoot. So walkers move to the edge of the path making it even wider.

- Trampling spreads because some of the moorland paths are poorly defined.

1 a) Identify the main physical and human factors responsible for footpath erosion.

 b) Construct a flow chart linking the factors to the processes of footpath erosion.

NGOs working in the Shropshire Hills

The National Trust is a charity and just one example of an NGO working in the Shropshire Hills. Its aim is to conserve the countryside and country houses. It owns two particularly significant landscape features within the Shropshire Hills:

- Long Mynd (Figure 5.5): a large upland plateau with a steep western scarp slope and steeply incised gullies on the east. Some of these gullies contain unusual flora and are notified SSSIs. Another larger gully, Cardingmill Valley, is a honey pot site that attracts thousands of visitors. The National Trust has a visitors' centre and education officer here to deal with the many visitors and parties of school children.
- Wenlock Edge: a wooded limestone escarpment which has exceptional wildlife interest. The National Trust has purchased a continuous 10-kilometre length of the ridge, improved footpath provision and provided a wheelchair route.

▲ **Figure 5.5** The wild landscape of the Long Mynd. The deep gullies, known locally as 'batches', were cut by glacial meltwater at the end of the last ice age.

▼ **Figure 5.6** From 'Long Mynd Access Map', The National Trust, 1998.

Access by CAR

Please use the car parks
If you're arriving by car, please park in the Carding Mill Valley or in Church Stretton. The Carding Mill Valley has three car parks, and is open all year round. It also has the facilities you need, including lavatories, disabled access, tearoom, shop and an excellent network of walking routes accessible directly from the main car parks.

There is very limited parking in or around any of the other valleys, so please plan all your walking or cycling around starting points in Carding Mill Valley or Church Stretton.

Don't go over the top!
Roads accross the hill – particularly the Burway – are very narrow and difficult. They can be very congested in summer, and often impassable in Winter. Please don't try to drive over the hill unless it's absolutely necessary.

Keep on the road
There is no off-road driving for unauthorised vehicles anywhere on the hill – 4x4 vehicles can cause serious damage to heather, ground nesting birds and to grazing. You may occasionally see vehicle tracks, but these can only be used by farmers for looking after sheep in emergencies, and generally don't lead anywhere. Access along these routes is for authorised vehicles only.

Think green!
Next time you visit, please think about using public transport! Cutting down on car usage is something we can all do to help protect areas like the Shropshire Hills from congestion, fumes and the effects of greenhouse gases.

1 Study Figure 5.6.
  a) What conflicts are the National Trust trying to avoid?
  b) What do the cartoons suggest about access to the Long Mynd?

The Shropshire Hills Joint Advisory Committee

The Shropshire Hills JAC first met in 1993. Its initial membership is listed in Figure 5.8. The first meetings of the JAC aimed to identify the main issues facing the Shropshire Hills. From this it produced its first Management Plan: the Shropshire Hills Advisory Plan 1996–2006. The Advisory Plan sees its main aims as being to:

- conserve the unique and distinctive qualities of the area
- monitor changes which may have a serious impact on the landscape
- encourage enjoyment of the countryside
- contribute to the long-term future of the Shropshire Hills by promoting sustainable developments (see page 74).

The plan has objectives which cover specific issues such as landscape, agriculture, forestry and nature conservation. The issues relating to tourism and recreation are listed in Figure 5.9. Reading this will help you to decide the objectives that *you* would set if you were involved in the JAC.

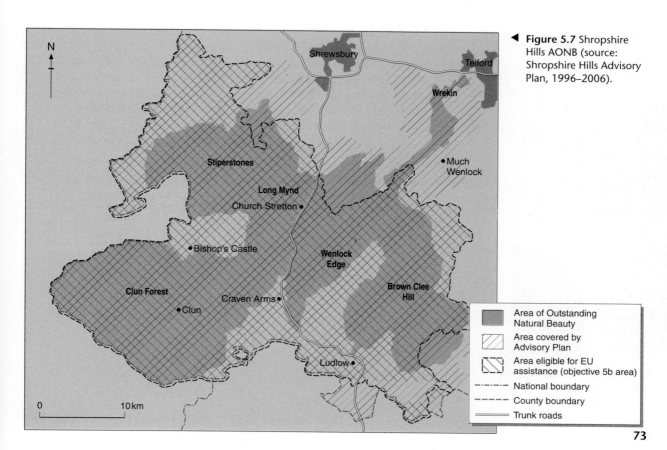

Figure 5.7 Shropshire Hills AONB (source: Shropshire Hills Advisory Plan, 1996–2006).

▼ **Figure 5.8** The members of the Shropshire Hills Advisory Committee (source: Shropshire Hills Advisory Plan 1996–2006, Shropshire Hills Joint Advisory Committee).

Local authorities
Shropshire County Council
Bridgenorth District Council
Shrewsbury and Atcham Borough Council
South Shropshire District Council
Wrekin District Council
Association of Town and Parish Councils
Statutory/voluntary agencies
British Horse Society
British Trust for Conservation Volunteers
Community Council of Shropshire
Countryside Commission
Country Landowners' Association
Council for the Protection of Rural England
English Heritage
English Nature

Farming and Wildlife Advisory Group
Forestry Authority
Forest Enterprise
Friends of the Earth
Institute of Chartered Foresters
Ministry of Agriculture Fisheries and Food
National Farmers Union
National Rivers Authority (now Environment Agency)
National Trust
Ramblers' Association
Rural Development Commission
Shropshire Countryside Liaison Committee
Shropshire Wildlife Trust
Shropshire Hills Countryside Unit
Timber Growers Association
The Sports Council

▶ **Figure 5.9** From Shropshire Hills Advisory Plan 1996–2006 (source: Shropshire Hills Joint Advisory Committee).

1 Take each of the issues a-f in Figure 5.9. Identify which organizations in Figure 5.8 may be involved in each issue.

2 How far do you think the JAC is simply a collection of pressure groups representing a single interest?

Issues relating to tourism and recreation:

a) Developing sustainable tourism in a way that does not compromise the qualities of the Shropshire Hills.

b) Identifying recreational use of the Shropshire Hills so that organisations involved in the provision of recreation can work towards ensuring that the area is protected from over-use.

c) The impact and appropriateness of recreation activities need to be considered.

d) How to co-ordinate information on the Shropshire Hills, especially to ensure that the message that it is a special (and sometimes fragile) landscape is communicated.

e) Encouraging visitors to the countryside by car raises a number of issues such as parking and pollution.

f) Some concerns have been raised over themed walks in the Shropshire Hills, which are often linked to holiday packages. Greater awareness locally of the development of such walks and agreement of positioning of waymarks needs to be addressed.

▼ **Figure 5.10** Primary visitor activities in Shropshire Hills (%) (source: 'A tourism strategy and action programme for South Shropshire', Shropshire Sustainable Rural Tourism Project).

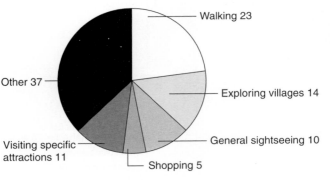

▼ **Figure 5.11** Mode of transport of visitors to Shropshire Hills (%) (source: 'A tourism strategy and action programme for South Shropshire', Shropshire Sustainable Rural Tourism Project).

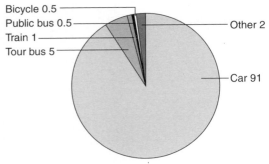

▼ **Figure 5.12** Main appeal of Shropshire Hills to visitors – some respondents indicated more than one (%) (source: 'A tourism strategy and action programme for South Shropshire', Shropshire Sustainable Rural Tourism Project).

Countryside and hills	48
Scenic/beauty	35
Peace and quiet/relaxing atmosphere	13
Buildings/architecture	12

▼ **Figure 5.13** Main dislikes of visitors to Shropshire Hills (%) (source: 'A tourism strategy and action programme for South Shropshire', Shropshire Sustainable Rural Tourism Project).

Inadequate signposting	7
Lack of adequate interpretation	4
Poor toilets	4
Inadequacy of waymarking signage	3
Lack and expense of car parking	2
Too many tourists	0.1

You are on a two-week placement with the Shropshire Hills Joint Advisory Committee. As an A–level student, the committee has asked you to produce a report on how well the Shropshire Hills are managed, and how it might encourage local initiatives to promote a more sustainable economy and environment. In your report:

a) assess the value of the landscape of South Shropshire.

b) identify the main tourist 'hot spots'.

c) show how far tourism in the area *is* or *is not* sustainable, using examples of benefits and problems brought to the area.

d) explain your ideas and suggestions for addressing each of the issues a-f in Figure 5.9 sustainably, and who might be responsible for making sure that they can happen.

Ideas for further study

Investigate the development of rural leisure and tourism in a rural area close to your school or college. What leisure opportunities currently exist in the rural area? What impacts does leisure activity have on the environment, economy and society of the rural area. How might leisure provision be extended in the future and what impacts would this have?

Summary

Job losses in traditional rural occupations and the spread of counterurbanization threaten to make rural villages little more than dormitories for commuters. Rural communities need new work opportunities if they are to survive. The rural economy is, therefore, diversifying. The leisure and tourism industry offers new jobs in some rural regions. Such economic developments will be of most benefit to the rural community and environment if they are sustainable. Decision making about the rural economy is made by a wide range of individuals, statutory bodies and NGOs. These decision makers often act in co-operative groups or consortia.

References and further reading

The impact of leisure and recreation in rural areas of the UK are discussed in:

Robert Prosser (1994) *Leisure, recreation and tourism.* Collins Educational. (especially chapters 7 and 8).

Susan Owens and Peter L Owens (1994) *Environment, resources and conservation.* Cambridge University Press. (especially chapter 6).

6 Writing an A level geography essay

This chapter examines the stages of essay writing. As well as working through these in turn, you will need to develop good communication skills. These skills include the use of:

- correct language and geographical terminology
- good grammar and punctuation
- a detached and objective style
- accurate, labelled sketch maps and diagrams.

1 Researching your essay

The research process may take several hours and span a large part of the course. Research means using books, magazines, newspapers, CDs or the internet to develop your knowledge. Your understanding of the topic will gradually improve as you do this. You will need to note down key data and examples as these are the evidence that you will use to support your ideas.

Develop the use of geographical or correct terminology as you research. Avoid general terms such as 'scenic' when AONB may be more accurate, or 'tourists' generally when you may need to refer separately to day visitors, long-stay visitors or holiday home owners.

2 Read the essay title

Many students misread essay titles, identifying just one word in the title and writing everything they know about it. Answers like this rarely achieve a high mark. 'A' level essay titles are often complex and contain several key words. Focus on these key words. Study Figure 6.1 in which the key words of a title have been annotated.

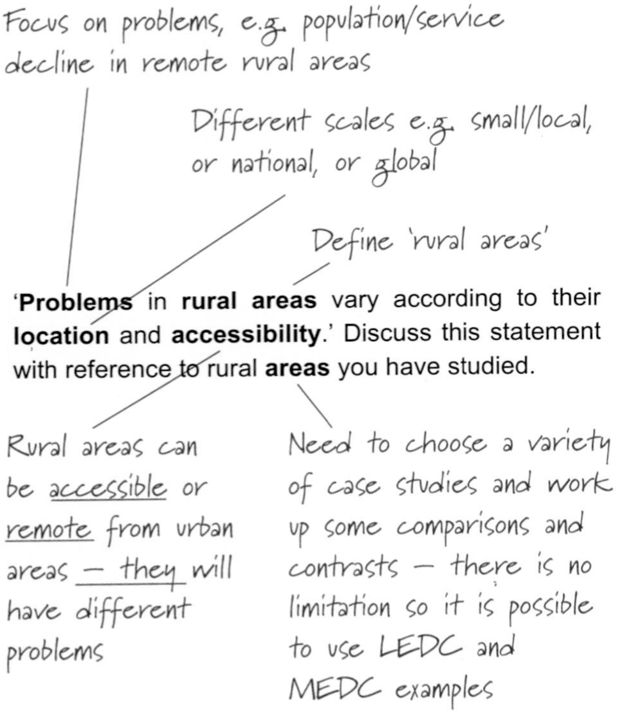

3 Planning the essay

Before you start writing, plan! In an exam you will have 40–60 minutes to write an essay. The first 5–10 minutes should be spent planning.

A plan is a sketch of all the ideas that you want to develop. It should be in note form rather than continuous prose and can be a list or spider

▼ **Figure 6.1** Focus on the key words of the title.

Focus on problems, e.g. population/service decline in remote rural areas

Different scales e.g. small/local, or national, or global

Define 'rural areas'

'**Problems** in **rural areas** vary according to their **location** and **accessibility**.' Discuss this statement with reference to rural **areas** you have studied.

Rural areas can be accessible or remote from urban areas — they will have different problems

Need to choose a variety of case studies and work up some comparisons and contrasts — there is no limitation so it is possible to use LEDC and MEDC examples

diagram (Figure 6.2). Its purpose is to identify each section or argument that you will develop. The bubbles in Figure 6.2 each contain a factor which influences the management of rural areas. Each bubble becomes a separate paragraph of the essay. Some ideas have already been added to show how the factors could be elaborated. This kind of plan will prevent you from getting sidetracked or forgetting to include important information. As you plan, try to work out what your argument will be: do rural problems arise from the location and accessability of rural areas.

Study the mark scheme in Figure 6.3. In order to score well in the section headed 'Skills' your essay must have a 'coherent line of argument showing evidence of planning and a logical structure.' This means that a good essay will flow from one idea to the next.

Identify the main features of rural environments which make them difficult to manage. You should illustrate your answer with reference to a variety of rural environments you have studied.

▼ **Figure 6.2** A spider diagram essay plan.

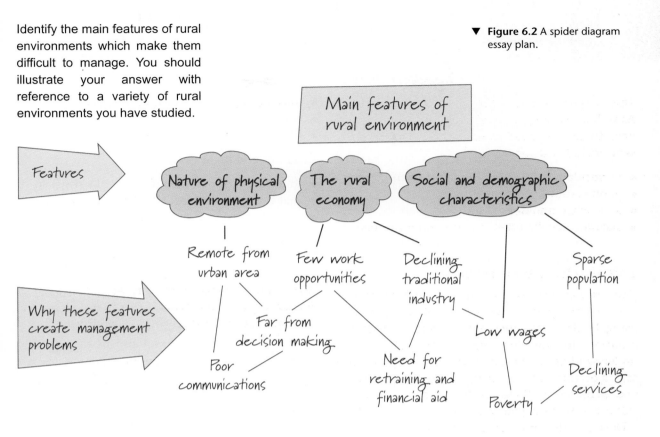

▼ **Figure 6.3** Essay marking scheme (EdExcel, January 1998).

Knowledge of concepts, issues and case studies:

8–7	Sound knowledge of case studies, concepts and issues; relevant and detailed case study(ies) information showing sound locational knowledge.
6–4	Satisfactory knowledge of case studies, issues and concepts; may lack location or precise detail and not always relevant selection; may lack depth and/or range at lower end.
3–1	Generalised knowledge with a lack of specific reference to case study(ies); knowledge of concepts and issues less clear and may be inappropriate to the question – considerable irrelevance.

Understanding of issues and concepts and their application:

10–7	Sound understanding of concepts, well applied case study(ies) with relevance to the answer, shows evidence of evaluative comment.
6–4	Satisfactory understanding with some relevant application of case study material; less evidence of ability to evaluate and knowledge not as well applied to understanding of the question. At lower end tends to be descriptive.
3–1	Weak understanding of concepts and issues with limited reference to relevant case study material. Focus descriptive rather than evaluative.

Skills:

7–6	Good use of language including appropriate geographical terminology. Coherent line of relevant argument showing evidence of planning and a logical structure. Possible use of relevant diagrams and cartographic techniques. Clarity of expression with high standard of accuracy in the use of punctuation, grammar and spelling.
5–4	Reasonable clarity and fluency of expression with some use of appropriate geographical terminology. Arguments are generally relevant with some evidence of planning and structure. Some use of illustrative material, accuracy in the use of punctuation, grammar and spelling.
3–1	Satisfactory clarity of expression of basic ideas but not always in a logical structure. Limited use of A Level geographical terminology with little evidence of planning and argument. Basic use of English but with mistakes.

1   Study the following essay title:
'"Community based rural development programmes are invariably the most successful." Assess the truth of this statement with reference to programmes you have studied in both MEDCs and LEDCs.'

a) Make a copy of this essay title and annotate it to show the main focus of the title.

b) Draw a spider diagram essay plan for this title. Try to include three or four main bubbles that will form the main focus for each paragraph of the essay.

4 Writing an introduction

The introduction of an essay is short – about 200 words – and should take about five minutes. Most introductions have two main aims:

- to define briefly any concepts or terms used in the essay title
- to outline the main ideas that you will develop in your essay.

Use Figure 6.1 to help you write an introducton.

5 Developing each argument

This is the main part of your essay and it will take about 25–35 minutes. Each new idea or argument should have its own paragraph and should be supported by evidence from your case studies.

Refer again to the mark scheme in Figure 6.3. To achieve a high mark, case studies must:

- 'be detailed' and show 'sound locational knowledge'. This means that you must **name** and **locate** the case study as carefully as possible. You must also include some detail, e.g. facts and figures. Use labelled sketches and maps to help you.
- be 'well applied' and show 'relevance to the answer'. This means you must choose case studies carefully to suit the focus of the title. Where possible, use case studies from your own fieldwork and/or coursework. Again, use maps or diagrams to help you.

When developing an idea you should avoid vague or general comments. You must try to convince the reader by using detailed descriptions and explanations in your evidence.

- quantify – add some facts and figures
- exemplify – give an example of a place
- illustrate – using good diagrams or sketch maps.

Diagrams and maps need:

- to be large enough (about $\frac{1}{4}$ – $\frac{1}{3}$ of a page) so that they are clear and visually effective
- to be detailed and labelled (known as annotated)
- a heading.

1   Study the essay title in Figure 6.1 again. Identify one concept and one case study that you could use in your answer. Draw a diagram of your concept and an annotated sketch map of your case study.

6 Writing a conclusion

Your conclusion, which in an exam will take 5–10 minutes, should draw your argument together. It is an opportunity to assess the relative importance of each piece of the jigsaw that you have been building in your essay. For example, the plan in Figure 6.2 identifies three main features that make rural environments difficult to manage. The conclusion is an opportunity to identify which of these factors is the most important.

Managing Rural Environments: Summary

	Key questions	Ideas and concepts	Examples used in this book
Definitions	• What do we mean by rural areas? • How and why do rural areas vary in landscape and character?	1 Rural areas may be defined by population, landscape characteristics, or using socio-economic data. 2 Rural areas vary on a scale, from urban fringes to landscapes of remote areas.	• South Shropshire (local and regional scale) • Machakos (Kenya), Andhra Pradesh (India) (regional scale)

Issues and processes	• What are the pressures on rural areas? How and why are pressures on such areas increasing?
	• What are the social, economic and environmental issues facing rural areas?
	• What are the causes of these issues?
		1 **Social**
		  a) Issues result from population change – decline and expansion. These issues may vary between MEDCs and LEDCs.
		  b) New developments in rural areas may conflict with traditional ways of life.
		2 **Economic** – Employment opportunity and service provision may be lacking in rural areas, resulting in relative poverty.
		3 **Environmental** – Changes in economic activity, such as farming and tourism, may result in environmental pressures.
			Social and economic
			• Bishop's Castle (*local scale*), Machakos, Kenya, Andhra Pradesh (India), South Shropshire (UK), Andalucia (Spain) (*regional scale*), rural-urban migration in Africa (*national and international scale*).
			Environmental
			• The effects of the CAP and regional aid in the EU. Agroforestry in Siaya and South Nyanza (Kenya).

Managing rural areas	• How can the pressures on rural areas be managed? • Who are the main agencies involved in managing rural areas, and how effective is their role? • What are the possible futures for rural areas?	1 Different agencies operate at different levels in managing rural areas; these may be public, private, voluntary, or a mix of these. 2 Different strategies may be needed to resolve social, economic and environmental issues and problems which face rural areas. 3 Some strategies may be socially, economically and environmentally suitable, others less so.	Agencies responsible for management: • South Shropshire, UK – local or regional councils, agencies such as the Countryside Commission, national or EU government. • Siaya, South Nyanza, and Machakos (Kenya) and Andhra Pradesh (India) – the role of NGOs or voluntary organisations.

Index